Praise
COMPANY S

"Bobby Leonard's *Company Secrets* has something for everyone: action, suspense, mystery, murder, romance...not to mention enough technical detail to satisfy even the aerospace engineers and metallurgists in the crowd. And it's a real page-turner! I found myself getting caught up in the drama and suspense of the story, especially with Dan's and Mary's adventures. I kept reading to find out what was going to happen next. And there was more than one surprise along the way. Given my background in engineering, the military and my travels to India and the Middle East, I could appreciate the technical and geographic detail and how well researched the book was. Surely Dan and Mary will have more interesting consulting gigs in the future that would give Bobby Leonard enough material for a sequel. I hope so!"

—MICHAL ROBINSON. Engineer, U.S. Army, consultant

"*Company Secrets* by Bobby Leonard will capture your interest from the "get go." Besides the action and the technical information, Leonard details the need of business consultants in critical situations. Leonard also specifically addresses the "whats", the "whys", for consultants in the space industry. Let's hope that there is a movie in the offing."

—MITCH IADAROLA. Retired U.S. Army Lt. Colonel

"Bobby Leonard's book, *Company Secrets*, will keep readers interest and it has all the ingredients a mystery needs from start to finish: suspense, murder, and romance. This page turner features surprises galore. Readers will be unable to put the book down. And there is nice balance of technical stuff in the book. Here's hoping this is not the last book we'll see by Bobby Leonard."

—JAMES CAPELLA. Retired State of Connecticut Inspector

"*Company Secrets* is packed with interesting characters who provide the fodder for this carefully crafted story. The book weaves its way from coast to coast, internationally, and back seamlessly. Readers will wait in anticipation for what comes next: murder, espionage, intimacy, lying, and enough mystery to keep one guessing who is doing what. Technically the book is very informative and keeps one up late turning the pages. Please, Bobby Leonard, more stories."

—RICHARD HAWKYARD. Retired small business owner

COMPANY
SECRETS

A NOVEL

Michelle,
Great seeing you and
Bill. Please enjoy
This Book all The way
Through!

Bobby Leonard

COMPANY SECRETS

A NOVEL

Bobby Leonard

McCaa Books • Santa Rosa

McCaa Books
1604 Deer Run
Santa Rosa, CA 95405-7535

This book is a work of fiction. References to real people, events,
establishments, organizations, or locales are intended only to provide a
sense of authenticity, and are used fictitiously. All other characters, and
all incidents and dialogue, are drawn from the author's imagination
and are not to be construed as real.

First published in 2018 by McCaa Books,
an imprint of McCaa Publications.

Library of Congress Control Number: 2018957231
ISBN 978-0-9996956-6-1

Printed in the United States of America
Set in Minion Pro
Author's photograph by Sue Dropp Portrait Design

www.mccaabooks.com

Prologue

Somewhere near Kandahar, Afghanistan

Four hours of blistering heat and dust rose from the dirt road near Kandahar. The Fixer, as his clients called him, was driving to meet Abdul Wazir in the southeastern hills.

"I don't know this Abdul, but my errand will cost him more than my usual fee because he demanded I drop everything and come to the mine," Fixer said to himself. "Ah, my fee? I'll wait to see what this job is."

Arriving at the site, he saw the three mobile trailers and headed straight to the middle trailer to see Abdul. Abdul met him at the door and introduced himself.

"I am Poya," Fixer said. "What goes on here?"

"I get mining equipment for operations. Can't tell you what's mined here. It's secret," said Abdul, pausing to watch Poya's expression. "You can't tell anyone about me, this mine, this location, or any other location I sent you. Agree?"

"You don't know who I am," said Poya. "I not just errand boy. I am Mohammad Poya. I am the Fixer, well known in Kandahar for doing important jobs for American army. I am Sunni Islam, and have friends with the Shias. I assassinated warlords before. I survived hard time when Taliban ruled most of Afghanistan. I make trust over years. I even worked for Hamid Karzai. I listened to American colonels and delivered messages to warlords and Taliban. What work do you have for me?"

"Pick up packages from places and deliver to me. Deliver packages from here to places in Kandahar. Some places may be shipping companies. I demand you be quiet and you use connections you have. I pay good for you; be fast and tell no one. Afgani AFN 30,000 for each package. I have no U.S. dollars to pay, but this equal U.S. $440."

"You want me pick up and deliver items between here and Kandahar and not tell anyone. What connections do you need from shipper?" asked Poya.

"Make sure each package gets no special attention from the inspectors. You do that?"

"Yes, for surcharge," said Poya.

"If you ask them not to look, makes them more suspicious. You tell them story that contents are samples, if they ask. I don't want suspicion," said Abdul. "Can you do that?"

"What are contents?" asked Poya.

"Secret material. Won't hurt people. I cannot risk having them get in competitors' hands. You understand?" asked Abdul.

Poya looked around at the landscape of dirt and rocks. "What is special here?"

Abdul ignored the question. "Also we talk on my special phone—both text and voice. You must come when I call. I tell you time to pick up package and where to deliver."

"I do this, but I need AFN 200 more. How many deliveries do you need, for how long?" asked Poya.

"Four deliveries a week for six weeks. Later, I need two or four per month. We done in six months," said Abdul.

Poya did a little math in his head and estimated his future earnings. That would be more than ten years income in Afghanistan.

"We have deal," said Poya. "Good. I tell you more later so you learn how I work with you."

On his drive back to Kandahar, Poya took his time driving the dirt road. The sun was setting in front of him, casting a reddish glow over the countryside. He felt good about the project and the prospect of making money with little risk in the coming months.

"I thank my friend for giving my name," he said to himself as he looked into his rear view mirror. On the North Slope he could see the top of a mine platform.

1

New York, New York

"THIS EXECUTIVE SESSION OF THE BOARD OF DIRECTORS is called to order on this day, Thursday, January 8th, at 3:00 p.m.," recited Rajah Malani, Chairman of the Board for AeroStar Corporation. "Secretary, will you please take role call." He banged his gavel as he always had in the past, and then turned the floor over to the Secretary.

"Please recite 'present' when your name is called," said the Board Secretary.

> Board of Directors: Rajah Malani, President and CEO of AeroStar; Torri Berri, President of ComStar; Harold Zaben, President of Vega Group; and John Wright, President of Wright Industries.
>
> Outside directors: Andrew Johnson, Roxanne M. Runnels, Kenneth S. Dubois, Linda A. Cook, H. M. McIntyre, and Richard Smith.
>
> Invited guests: Jesse White, Executive Vice President, Business Development & Strategy of AeroStar; Parker Jones, Senior Vice President and Executive Sponsor of Aether Program; Wade Williams, Senior Manager and Project Director of Aether Program.

" . . . And that completes our role call," said the Board Secretary. "Mr. Chairman, you may proceed."

This is the tenth director's meeting Rajah has chaired after being appointed CEO and Chairman of the Board of Directors for AeroStar Corporation. He's comfortable in this role after more than thirty-three years of service with the company. He's proud of the fact that he started his career in the aero design lab at ComStar in St Louis, Missouri, in 1982. He was just twenty-five years old with a B.S. degree in Aeronautical Engineering from UC Berkley, and a Ph.D. from Cal Tech in Pasadena, California. He also earned his MBA at nights in his first three years at ComStar.

This is no ordinary board meeting; that's why he has called for an "Executive Session." Rajah knows that securing the board's approval to proceed with their latest strategic initiative, Aether Program, requires cooperation from his executive staff and evidence that the strategic investment is sound.

Rajah is using the Executive Session format to protect the strategic initiative from being published to the shareholders. He's confident that the technology they will be presenting is sound and the results will revolutionize space travel. The rewards will allow AeroStar to grow significantly over the next twenty-five years, and the business growth will allow all the divisions to benefit significantly through integrated synergies that will contribute to the total initiative.

The company is sitting on breakthrough technology worth billions. He must keep this information secret to prevent the competition in the space technology business at bay. Today he will gain the Board's approval to go forward.

"Thank you," said Rajah. "Let's move to the first agenda item: Aether Program. Last year The Board approved preliminary funding to complete a cost/benefit analysis to determine the feasibility of entering the space launch and delivery marketplace. NASA retired the space shuttle program in 2011 and is launching its Advance Space Transportation Program to privatize the design and delivery of space vehicles for future space exploration. Does anyone have any questions about NASA's plans?" Rajah paused for questions.

"SpaceX, our competition, was formed and has secured numerous contracts with NASA. Their Heavy Falcon project is currently under development and will deliver large payloads to orbit and up to the outer reaches of space," said Rajah with clear diction and only a hint of a Middle-Eastern accent.

Born in the United States, Rajah is a first-generation Hindu from India. His parents wanted their children to grow up American first, while keeping their Hindu religion and their family culture intact. Listening to him speak, you would never realize his Indian heritage. His dark skin and hair are the only visible signs of his heritage.

Rajah continued to the seven points in the report, "Let me reiterate why we are asking for your support to fund this strategic initiative:

1. SpaceX multi-staged rocket platforms are costly. Current cost to deliver supplies to the space station is still about $10,000 per pound of payload.

2. AeroStar is in a unique position to take advantage of its new technology in aircraft transport, propulsion, and spacecraft. The Aether Program has a team of the brightest engineers from ComStar, Wright Industrial, and the Vega Group working to complete a design that will be competitive with SpaceX.

3. We have more than ten years of research building plasma rocket motors for future space travel. FutureDyne Rocket Company's Deep Space Rocket (DSR) motor is being designed to replace conventionally fueled rockets. The DSR motor is a true nuclear-electric plasma propulsion system that will produce more than seven times more thrust than conventional rockets. Thrust capable of launching a single-stage-to-orbit spacecraft to earth orbit and beyond. This technology alone will revolutionize space travel.

4. Our spacecraft will use a 'low-energy' launch profile that will take off and land from runways

anywhere in the world. Using low-energy thrust from conventional jet engines allows the vehicle to circle the earth gaining speed and altitude. The DSR takes over boosting the spacecraft to high earth orbit or deep outer space travel. Our DSR motor is revolutionary, giving the thrust necessary to achieve space travel and then return to earth. Our spacecraft will enter earth's atmosphere and fly to a landing site like the space shuttle. The space shuttle had only one chance for landing. Ours is capable of executing a go-around maneuver if needed.

5. Our spacecraft and services will also help reduce the total cost of payload delivery. We believe the delivery cost to be in the range from $50 to $100 per pound of payload.

6. We've projected the cost to design and build prototype systems to be $15 billion. We need to fund the first $5 billion to secure a contract with NASA. This means we have to show working prototypes of the spacecraft and the DSR propulsion system. We believe the future contracts, beyond the initial $5 billion, total about $10 billion to fund through delivery, with an additional $2 to $3 billion per year for up to fifteen years for operations and support.

7. Our payback for the $5 billion will be three to four years. The net present value is $15.6 billion, and our internal rate of return is 35 percent.

8. Our technology advances give us a ten- to fifteen-year head start over SpaceX and any other competitor allowing us to achieve the superior ROI."

Rajah looked around before proceeding.

"Are there any questions about what we've completed with the initial funding of the cost/benefit and feasibility study?" asked Rajah, pausing to see if there are any questions.

"The Chair recognizes Mr. Andrew Johnson," said Rajah.

"Why do you think NASA will change its current launch and supply strategies to adopt our low-energy launch vehicle? Won't they want us to demonstrate our technology before they give us the go-ahead contracts?" asked Mr. Johnson.

"Mr. Zaben, will you please take this question?" asked Rajah.

"Gladly," said Harold Zaben, President of the Vega Group. "We successfully worked with NASA in the past to deliver eighteen Luna 1 launch vehicles landing the first man on the moon. We were the prime contractor, and to this day we are still highly regarded and respected because of our work and reputation." He paused. "The reason NASA will fund our projects is because of the projected 'very low' payload cost that only we can deliver. The initial contracts will demonstrate 'proof-of-concept.' Driving down costs fits perfectly into NASA's long-term strategy."

"The Chair recognizes Ms. Linda Cook," said Rajah.

"But you are still asking for $5 billion up front? Did I hear that correctly?" asked Ms. Cook. "And then you expect to have NASA fund the $10 billion to complete the rest of the development project. This is a significant portion of NASA's budget over the next three to five years. How do you expect to sell them on our delivery system?"

"I'll address Ms. Cook's question," said Jesse White, EVP Business Development & Strategy for AeroStar, and a long-time "good ole boy" from the South. "NASA is privatizing space travel because they want the best solution at the lowest cost. SpaceX believes they can achieve $500 per pound in twenty years. We doubt they can do it using their current technology. We've studied low-energy launch systems for a long time. There just hasn't been a rocket powerful enough. Our new rocket technology is the missing piece." He looked her in the face before continuing. "Our scientists discovered how to make a nuclear generator with enough electricity to run an ion thruster. This thruster is seven times more powerful than any rocket motor available today. My last point is that our spacecraft is reusable. NASA likes reusable vehicles. Much like the space shuttle, but with much less maintenance because we won't need to replace tiles. Once we show them our prototype, NASA will gladly fund our follow-on program."

"How do you know our technology will work?" asked Ms. Cook.

Jesse White looked directly at Parker Jones, head of the Aether Program. They had planned this response. "I'd like Mr. Jones to answer your question."

"We've been testing aircraft and spacecraft designs for more than fifteen years," said Parker. "Our leading-edge material technology is the secret to the success of our new DSR motor."

"Thank you, Mr. Jones," said Rajah as he took the lead. "Does anyone on the Board have additional questions before we move to vote on the funding?"

"Yes, I have a question for the team," said H. M. McIntyre. "I see in your proposal that once the demonstration phases are complete, the big picture is to manufacture up to fifteen vehicles with support and spare parts over ten years. Please explain your supply-chain sources for exotic materials used in the DSR? Will you have adequate supply? What international and political barriers might there be in securing enough materials to manufacture the planned fleet?"

Rajah ever so slightly shifted in his seat as he thought about the question, "You all know that we have minority ownership in Annokkha Drat Exports located in India. They mine and export rare metals like neodymium and samarium, instrumental in plasma rocketry and electrical components used in Vega Group's products. They also mine gadolinium, a metal instrumental in our nuclear generator."

"Where are these mines, and do we have adequate reserves to support the projected production?" asked Mr. McIntyre.

2

Park Avenue, New York

Dan Duggan's alarm woke him at 6 a.m. with a start. He remembered that he was going to his office in New York City, so he quickly showered and shaved. Looking in his closet, he selected a light gray two-piece suit from a selection of about a dozen, "I'll be ultra conservative today; besides, the light gray will keep my spirits high." He selected a white shirt with fine blue pinstripes, and a tie with red, blue and gray strips to complete the look and feel he was trying to achieve. No power suit today. He looked at his clean shaven face and brown hair—not too short, but neatly combed to the right side with just enough gel to keep it from falling in his face. His sideburns were trimmed neatly at the ears and the back expertly layered. He put the tie around his neck and pinched the overlap while he sized lengths. His favorite knot was the double Windsor because of the perfectly symmetrical knot. He synched it up and checked the length to ensure that it just touched the top of his black leather belt before running the other end of the tie through the maker's label. Dan seldom wore a tie clasp. He put on his suit jacket, closed one button, and then stepped back to see the whole ensemble including his black, cap-toe shoes. *Voila*, a perfectly dressed management consultant.

Dan lived in Stamford, Connecticut, only a few minutes from the Metro station via Uber. When he got on the train, he was still in a sour mood because his boss, Robert Kavanah, insisted that he stop everything and come to the office about a new engagement. He was in the middle of writing the final client engagement report. Robert had occasionally asked him to jump through hoops, insisting he drop everything. Well, he's the boss.

He rode the Metro-North train from Connecticut, his brain still wrapped around his current engagement. During the thirty-minute ride he read his email and project reports from the team. When the train lurched to a stop, he closed his laptop and started his ten-minute walk through mobs of people to the office. A warm blast of air hit him as it always does when he exited the train. The tunnel was smelly with steamy grease mixed with perfumes, aftershave, leather, and wet wool. He merged with hundreds of commuters and rushed to the tunnel exit, to the center of the terminal, and then up the escalator and through the MetLife building. Anderson & Smith's offices were just through the Park Avenue tunnel where it crosses East 46th Street. The air was cool and dry and helped him calm down from the summons he received.

Dan is a senior associate with Anderson & Smith and has done this commute from Connecticut for eleven months. Previously, he lived in New York City for five years. Most of the time he worked from home or at clients' sites. He's meeting with his partner about a new engagement that will run one to three months and require significant travel. Being a closer- finisher personality, he's irritated about this sudden change of plans. His friends think his job is glamorous—traveling to major metropolises around the world, staying at five-star hotels, eating at first-rate restaurants and bistros. They haven't walked in Dan's shoes working ten hours or more a day.

"What's up today?" Dan barked, entering Robert's office. The corner office had a clear view of Park Avenue, and this morning was so clear he could see all the way to Central Park and even make out the George Washington Bridge. The great view calmed him down for the meeting.

"Dan, we have an urgent request from my friend Rajah Malani, CEO of AeroStar." Robert is all business, no time for small talk. Not even a hello, how are you?

"He wants us to review their current business strategy. The Board wasn't sold on the plan he presented, and they asked for a second review before they approve funding of a new strategic initiative. He's beside himself and wants me to pull our best and brightest to conduct the assessment. I want you to lead this engagement because you have aerospace industry experience, and you're my top business strategy consultant."

"So I just drop my current project?" Dan asked with a little edge in his voice.

"Already cleared you to hand off your work to Allison Weldon. She'll finish the project and manage the team members. It's good for her to step up to take on more responsibility. I'll oversee the final report."

"Tell me more about AeroStar. What's wrong with the current business strategy? What issues did the Board have? And, where do we need to dig into the plan to satisfy them?" Dan tried to get to the heart of the engagement. He also displayed his own disappointment with his body language.

"Dan, ah, I'm sorry to do this to you again," said Robert. "If I could have put Rajah off, I would have, but his sense of urgency was very compelling. Besides, this is good money for us and more to come in the future." Robert paused to watch Dan's reaction. After what seemed like an eternity, Dan finally relaxed and smiled, the sour mood dissipated.

Robert continued, "But, before we start, you and your associate must sign a nondisclosure agreement. Then you can go to our library and review the annual report and other reports Rajah sent over. This is secret because of the sensitivity of their plan. None of the information we see and gather can ever be leaked or shared with anyone other than the top management team at AeroStar. You will work under a cover name for the project called 'process reengineering and supply chain management.' Your first step will be to meet with Rajah and his senior team to determine the scope and timing of the engagement and to write the engagement letter for his signature."

As Dan signed the nondisclosure agreement, Robert explained, "AeroStar started a strategic initiative—code name Aether—last year to investigate the feasibility of entering NASA's space race. They have new technology they believe will revolutionize space travel. The Board was not impressed with his presentation, at least not enough to fund $5 billion. The Board posed two questions he couldn't answer: 1. Will the technology work? 2. Do we have an adequate rare earth metals supply chain to produce all the spacecraft?"

"You need to get in and quickly gather the necessary information to assess their business strategy as it relates to their new technology. I suggest that you use our Issue-Based Technique so that we don't waste time," he continued. "I'm also assigning a new associate, Mary Johnson, to work with you. I know you'll do a good job mentoring her. She has excellent business strategy experience, but she hasn't used the Issue-Based Technique."

"I'll pull her curriculum vitae and work experience and begin preparing for my initial meeting with management."

Dan read the annual report and waited for Mary Johnson. He wondered why they named the project "Aether."

He typed on his laptop: What is Aether?

The computer screen displayed

Sky deity

Aether is the primeval god of the upper air . . .

I wonder who named the project, Aether, the primeval god of the upper air? Dan thought. He knew that aerospace firms love to use mythical names for their most secret projects. The names seem to come alive to fulfill their prophecy. This one must certainly fly to upper earth orbit or beyond. He wondered about NASA's business approach, knowing they retired the space shuttle. A new company, SpaceX, owned by a young man named Elon Musk, delivered supplies to the International Space Station (ISS). Mr. Musk also founded Tesla, a company that makes expensive electric cars. He even put his personal Tesla sedan on the nose cone of his Falcon 9 heavy rocket for its maiden flight. A dummy astronaut drove the car.

Dan typed: What is NASA's plan for space exploration?

The computer screen displayed

> Advanced Space Transportation Program: Paving the Highway to Space

Going to Mars, the stars, and beyond requires a vision for the future and innovative technology development to take us there. Scientists and engineers at NASA's Marshall Space Flight Center in Huntsville, Alabama, are paving the highway to space by developing technologies for 21st century space transportation.

As NASA's core technology program for all space transportation, the Advanced Space Transportation Program at the Marshall Center is pushing technologies that will dramatically increase the safety and reliability, and reduce the cost of space transportation. Today it costs \$10,000 to put a pound of payload in earth's orbit. NASA's goal is to reduce the cost of getting to space to hundreds of dollars per pound within twenty-five years and tens of dollars per pound within forty years…

NASA's plans sounded a little like a big corporation mission statement. They want to go to Mars and lower the cost of flying in space too. Dan printed the article and started a tab in the binder labeled Engagement Research Binder.

Now it's time to go look at what Elon Musk wants to do with his SpaceX Company.

Dan typed: What is SpaceX business plan?

The computer screen displayed

> Achievements of SpaceX include:
> - The first privately funded, liquid-fueled rocket (Falcon 1) to reach orbit (28 September 2008)
> - The first privately funded company to successfully launch (by Falcon 9) orbit and recover a spacecraft (Dragon) (9 December 2010)
> - The first private company to send a spacecraft (Dragon) to the International Space Station (25 May 2012)

- The first private company to send a satellite into geosynchronous orbit (SES-8, 3 December 2013)
- The landing of a first-stage orbital capable rocket (Falcon 9) (22 December 2015 1:40 UTC)

SpaceX Business Goals:

Musk has stated that one of his goals is to improve the cost and reliability of access to space, ultimately by a factor of ten. The company plans in 2004 called for "development of a heavy lift product and even a super-heavy, if there is customer demand." Each size increase would result in a significant decrease in cost per pound to orbit. <u>CEO Elon Musk said: "I believe $500 per pound ($1,100/kg) or less is achievable</u>."

Dan underlined his cost goal and put a copy in the engagement binder. He found it interesting that Elon Musk thinks he can lower costs also. That's about a twenty-fold decrease from current cost. Dan researched other topics to help refresh his aeronautical engineering knowledge and added them to the binder.

Dan was puzzled about the material technology for the nuclear-based plasma rocket motor. He referred to the Board minutes to get the three "rare-earth metals."

Dan typed: neodymium, samarium, and gadolinium:

- Another chief use of neodymium is as a component in the alloys used to make high-strength neodymium magnets—powerful permanent magnets.
- The major commercial application of samarium is in samarium-cobalt magnets, which have permanent magnetization second only to neodymium magnets; however, samarium compounds can withstand significantly higher temperatures, above 700 °C (1292 °F), without losing their magnetic properties.
- Gadolinium as a metal or salt has exceptionally high absorption of neutrons and therefore is used for shielding in neutron radiography and in

nuclear reactors. The world gadolinium supply is
found in Afghanistan.

I'm not a nuclear chemist, but these metals seem to have properties that improve magnets. I think they must have something to
do with their breakthrough research, thought Dan

He scanned down to the end of the Board minutes and read
the last few lines, "Hmm, they've already approved a half-million
dollars for our engagement. I think we have plenty of funding to
complete the engagement."

- Motion/Second/Carries 10-Y 0-N—Torri Berri motion "to
approve up to $500,000 for outside consulting review of the cur-
rent business plan."

- Action Plan: Harold Zaben to request assay report from
Gholam Sharma, Managing Director of Annokkha Drat Exports.

- Rajah Malani: "Board tables motion to approve the $5 bil-
lion strategic initiative until mine assay report and the strategy
assessment report by third party consulting group is completed.
Board of Directors meeting is adjourned."

3

New York City

KEVIN KOUBIEL'S CONTRACT SALES BUSINESS he runs from his apartment in New York City changed eighteen months ago when he received a voice message on his home office answering machine that intrigued him. "My name is Frankie Gallo. I have a unique offer for you to join our team. Call me at area code 973-425-1575."

Kevin got on his computer and searched for the name Frankie Gallo. He learned that he runs a technology business in New Jersey—computer and software sales. He also read some New Jersey news articles purporting that Mr. Gallo is a New Jersey mobster specializing in technology-related scams. He noted that the writer referred to Mr. Gallo as Il Capo, his underworld nickname. Kevin jotted some notes on a pad and returned his call.

The number he dialed was answered immediately. "Frankie Gallo. And you are the famous Mr. Kevin Koubiel."

"Yes Il Capo, I got your message and I'm intrigued that you may have an opportunity for me. I run a legitimate business, so I don't know why you called."

"Listen to what I have to say this one time," said Frankie, aka Il Capo, in a low-toned voice with a heavy accent. "You know me because of my name and reputation. You don't know this. I started a new business. It's gonna be big, very big, and it is legal. Lithium, ever heard of it?" He pauses waiting for Kevin to reply—nothing. "This stuff is the latest thing. Ever heard of a lithium ion battery? It's big in electric cars. I formed a new company and I have a part-ner—a large mining company. They will mine the lithium. My company buys the mining leases in North America. I have a busi-ness plan too. Want you to look at it and you decide."

"I'm listening," said Kevin.

"We—I mean you—will quietly lease lithium mineral rights in America. No one has a plan like ours. We will become the leading owner of lithium mineral leases in America."

"Why do you want me? I know nothing about minerals or min-ing or lithium. I sell equipment and parts."

"Three reasons: one, you are good closing deals; two, you get the job done—no matter how complex; and three, you're organized, a hard worker. We have a lot of people to meet and deals to close. That's why I want you." He spoke with a low, extended youuuuuuu.

"I'm flattered at your confidence in me, but I'm really doing well with my current business."

"Here's the deal—I pay you a salary $150 g's, all expenses to cover travel, office, and other costs. I offer you 100,000 shares in our company with additional shares and options based on per-formance signing leases," said Frankie. "You sign our NDA, then I show you my financial plan. My accountant calls it a valuation model. It shows exactly how the stock value will grow the next few years. He estimates company will be worth upwards of $1 billion. You heard it, billion with a capital B!"

"Mr. Gallo, I don't want to insult you. You have my attention. I will look into your financial and business plans. I will treat this opportunity like I would any other—with respect and dignity. I would hope that you would also respect mine."

"Mr. Koubiel, trust me. This is real. I will send you plenty of research we've done. You can search the corporation documents also. The new company is U.S. Lithium Mines Corporation in Delaware, Wyoming, Utah, Colorado, and Nevada. You see our partner company is from India. Company is Annokkha Drat

Exports. I want you to know everything we know, but you can't tell anyone! Do you understand?"

Kevin signed the NDA and conducted his own research into lithium before he agreed to work for the company. He was equally impressed with the worldwide projections of lithium production over the next twenty years. Yes, it was a real opportunity and he would make sure that his contract was rock solid!

EIGHTEEN MONTHS LATER, Kevin Koubiel's operation in Rock Springs was moving according to plan. He was like the energizer bunny. He rented a small apartment on the west side of Rock Springs where he could work with anonymity. He was well organized and systematic in his plan to secure mineral leases in a four-state area—Wyoming, Utah, Nevada, and Colorado.

While he knew little about lithium and mining, he was an excellent salesman with an uncanny ability to gain people's trust. He started by researching the property records in the state, and then reaching out to each landowner. He used a clandestine approach, keeping everything secret. It worked because he was believable, and people want to make a lot of money.

Kevin knew that the current generation in Wyoming was ready to cash in. Why not? Oil and gas leases were making many land owners rich indeed. And now, lithium was really going to make a difference in the world too. And there was lots of lithium in the four states.

The operation worked by having Mr. Koubiel sign lithium mineral leases that excluded all other oil, gas, and minerals. To secure the lease, the Bureau of Land Management (BLM) required a signed and notarized lease and a recent assay showing that lithium had been discovered on the property. Kevin hired a local lawyer, held by nondisclosure agreement, to file the leases. The process was similar to filing any legal document at the courthouse; only the filing was done at the appropriate Bureaus of Land Management (BLM) office. It was similar to the way gold claims were staked in California during the gold rush.

Anyone can go to the BLM office and conduct searches to see what leases have been filed, who owned them, the property boundaries, the minerals included, and the assay results for the lease. Kevin spent some of his time finding land that had no lithium mineral leases. The pickings were really easy. He lined up the leases, and his office manager in Brooklyn filed the paperwork and maintained necessary documentation.

4

St. Louis, Missouri

Lights were out in the ComStar's design lab except for the dim glow shining from cube A-59. Across the aisle the printers were quiet; the large, high-tech 3-D printers looked like big mechanical arms stopped in strange positions, as if they were thinking about their next chess move.

A faint cell phone chime rang from the cubical. "Hello Bill," answered Dr. Summer Sexton. "What's up? Where are you so late this evening?" As she waited for a reply, she could hear buffeting sounds like wind rushing through sheets hung on a clothesline.

"I need an update from you," replied Bill with a heavy and slurred drawl. "The Board didn't approve our request; they want more proof that our designs will work."

"You should speak directly with Wyatt. He's my boss and I don't need you going around him."

"I wanted to hear your voice again. You just have a way of explaining the design to me and how it will work."

William "Bill" Anderson has been ComStar's VP of Research and Development for the past ten years and is a key manager in the Aether Program. He came up through the ranks during the heyday of the SST race in the 1970s and helped build the highly successful SW-ll commercial transport. It was the first commercial flying wing design. He knows nothing about designing a hypersonic spacecraft.

Back in the days when he was in charge, he kept abreast of all the design pieces of the airplane puzzle. He challenged everyone's work—draftsmen, structural engineers, systems engineers, and aerodynamics engineers—to test their conviction and understanding of their work. Everyone working for him paid attention. He expected excellence from his team then, as he does now. But today, he found it hard to keep up with the many computer design iterations, so he relied on frequent discussions with his key team members, Summer Sexton and Wyatt Calvert. He preferred to talk to Summer.

Bill has relied mostly on Summer's help to explain the technical nuances of the spacecraft design. Sophisticated computer programs tests many design alternatives. The design team selected hypersonic delta wing because it was the best fit for this mission. The drawback was its poor performance, flying slowly during take-off and landing. So they decided to add a swing-wing to improve slow flight modes.

"Bill, cut the crap," said Summer. "You shouldn't be talking to me, really. Why don't you call Wyatt in the morning."

"I'll be back in the office tomorrow. I've already asked Wyatt to brief me on the swing-wing design. I need to know exactly where you and Wyatt are in the design and what needs to be done next. This project is extremely important to this company, and all of us," said Bill.

The phone connection went silent when he hit the red button on his cell phone.

5

Rock Springs, Wyoming

FTERNOON BREEZES FELL OFF THE WESTERN PLATEAU, then
down across the valley, but the temperatures were perfect for
a championship rodeo. This is the last year that Rock Springs will
host the Annual National High School Finals Rodeo. The city of
Rock Springs has grown to 25,000 people mostly from coal mining,
minerals, and natural gas and oil shale production. Rock Springs
is doing well spending its money on facilities and amenities most
small towns could only dream about.

Life in Rock Springs was good; you could see the brand new
high-end trucks and off-road vehicles parked in most driveways.
Community parks dot both the new and old housing developments. Restaurants and fast-food joints are thriving. Over at the
Sweetwater Events Complex, the rodeo is in its third day.

"Ladies and gentlemen, please move your attention to the arena
and give our own local boy, Dale Butterfield, support as he mounts
his bareback bronco named Jumper," called the rodeo announcer.

As the gate swung back, Jumper left the chute, he bucked
straight out and up as Dale's left arm rose and his boots marked
the horse out of the bucking chute.

"What a great start," said the fancy-dressed cowboy watching the event from the grandstand. Eight seconds latter, Dale released his grip just as the bronco whipped him off the right side.

"So let's talk about why we are here," said Buck Jackson. Buck had received a call and decided to meet with the man, name unknown. He was intrigued with his caller because the man knew so much about his ranch, specifically the location and acres he owned. The caller knew that Buck had lost his wife a few years earlier to breast cancer. And, he mentioned that his employer wants to work with Buck. He could make significant income from a possible deal.

The man dressed in blue jeans, button-down shirt, and loafers nodded and said, "My employer is interested in the mineral rights on your property. Are you willing to make a deal? If you are, then what would you need from me to secure one?"

"Who do you work for?" asked Buck. "What minerals do they want?"

"In due time you will know who your partner will be, but we need to keep our discussions confidential. We want to understand your needs to ensure compatibility with my employer. You know, this is a two-way street—your needs to protect your property, and ours for the mineral. It's too early for lawyers until you and I have a solid meeting of the minds," said the unidentified man.

"I might be interested under certain conditions," said Buck. "First 'n foremost, I can't have you destroy property. My land has supported cattle, sheep, and horses for six generations, and we're not stopping I want money, or royalties, as you call them, for the rights and then a cut of minerals you mine. I just won't allow you to sit on the lease. Understand?"

"Yes, I think we can meet those needs," said the man. "I'll get some preliminary information from my employer, an offer so to speak. If you approve, then we'll put words to paper. But you must never talk about our discussion with anyone. Will you do that?"

"Look, I've lived my entire life in Wyoming. Inherited this ranch from my parents. My brother and sister also have original homesteads. Ranching is our whole life. When I lost Martha, I also lost a lot of motivation for ranching. I even considered selling. But

I told myself not to give in. Not for at least five years before making any decision."

"Next up is Skyler Holley from Columbus, Ohio," said the announcer.

"May I consult my lawyer?" asked Buck.

"Not yet, but in due time. Because of secrecy, my employer asks that you use this burner phone. We cannot be too careful. I'll call you next week to schedule our next meeting," said blue-jeaned man. "You may call me anytime."

"Why all the cloak and dagger?" asked Buck.

"It's quite simple. This is a strategic move for my employer, and any leak to the public could jeopardize their plans. Okay? Secrecy and timing are critical. We're not doing anything illegal. Okay? You most likely will get rich. I think you'll understand," said the man. "Also, I'll know if you go around asking questions about your mineral rights. That will kill the deal. Do you understand?"

"I can live with that for now," said Buck. "But I need to know your name. I need to know the man I'm doing business with. Do you understand?"

The unidentified man said, "Certainly. My name is Kevin Koubiel. And, you will get to know me a lot better."

A few minutes later, Buck and Kevin left the grandstand and went in separate directions.

6

Aerostar HQ, New York

DAN IS SEATED IN THE SECOND BOOTH at Mama Jo's on the corner of Park Avenue and East 47th waiting for Robert Kavanah. He is thinking about what to say at the meeting with AeroStar executives this morning. He has a number questions, but only wants to gather information to prepare a good engagement letter. Robert told him this engagement is "sole sourced" because of his relationship with Rajah. Price isn't an issue. Dan was deep in thought and didn't see Robert enter the diner.

"Good morning, Dan," said Robert.

"Morning. I've been thinking about how we should conduct this interview. You said we're meeting Parker Jones and his direct report, Wade Williams. Will there be anyone else at this meeting?"

"We will most likely meet Jesse White, Parker's boss. He must give his stamp of approval on our engagement letter."

They got their coffee, and then Dan asked, "How do you want to conduct this meeting?" Not waiting for an answer, Dan continued. "I think you should talk about your long-lasting relationship with AeroStar and Mr. Rajah Malani. Then, I will outline our qualifications and ask questions about their expectations."

"Excellent," said Robert. "Let's get started."

Dan looked at his notes from the preliminary information he reviewed, the Board minutes, and his research notes.

"I see that Ms. Cook wanted to know if the new technology will work," said Dan. "I believe she was referring to the new DSR motor that is powered by a nuclear generator which has never been built before. Then Mr. McIntyre questioned the supply of the rare earth metals needed to build the nuclear generator through full production of the spacecraft. We'll have to validate the supply of the rare earth metals."

"Good, let's start with these two questions," said Robert. "I'd like you to lead the questioning about these issues."

A few minutes later, they took a cab for the uptown ride to AeroStar's headquarters. Dan was comfortable with his pre-meeting plans. He will use a questioning technique called "Objective-Barrier-Question." This technique is a tool used by consultants to get to the heart of the clients business issues. The consultant first asks about the key objectives the client wants to achieve. Then asks to articulate what is preventing them from achieving each objective. The final thing the consultant does is pose potential questions in the form of hypothetical statements to explore potential ways to solve each problem. The end result helps the consultant formulate a clearer understanding of what the client needs to solve in the engagement.

Parker Jones greeted them and walked them into his office. Wade Williams was already sitting in the chair just to the right of Parker's desk. Dan took a careful look around Parker's office and noticed the walls were posted with lots of awards and pictures of Parker shaking hands with someone of importance. On his desk and credenza he saw a number of pictures of his wife and kids—hmm, a family man.

"Thank you for meeting with us on short notice," opened Parker. "I know Jesse White, our Executive Vice President of Strategy, called you about doing some business strategy work for us."

Robert Kavanah replied, "I have a long-lasting relationship with AeroStar and am pleased that you came to us about your strategic initiative."

Parker continued to break the ice by talking about the Yankees. "If you are a Met's fan, we won't hold that against you . . ." After a few looks and a little awkwardness, he said, "So let me tell you about our strategic initiative called the Aether Program. Last year the Board approved initial funding to develop a feasibility study based on new technology our company developed. This technology is revolutionary, giving spacecraft more capabilities, and would drastically lower the cost of space travel. Our Aether Program team is working diligently on work plans and designs for the new spacecraft. However, our Board needs a second opinion before making a significant investment decision."

"We see that your team has put their blood, sweat, and tears in to this feasibility study," said Robert Kavanah. "We're here to understand how we can help you get the Board's support. Will you tell us more about your expectations and timing for this strategy review you want us to perform?"

"I believe you were given a copy of our Board of Director meeting minutes where we answered all their questions," replied Parker. "Some of the board members need convincing that our technology will work. And one board member even questioned our sources of rare-earth metals. We weren't prepared to answer those kinds of questions."

Robert said, "Let's explore more about the details of your Aether Program so that we can prepare an engagement letter for you to approve. Dan is an experienced business strategy consultant and has direct industry experience in the aerospace industrial sector. We believe he's a perfect fit for AeroStar." Robert nodded to Dan to take over.

"I'd like to start by understanding some high-level company objectives or business goals, and then I'd like to drill into some of the specifics for the Aether Program," said Dan.

Parker and Wade looked at each other quizzically before nodding.

"What are the company's longer-term business goals that are supported by the Aether Program?" Dan asked.

"AeroStar wants to become a leading supplier to NASA for spacecraft, rocket motors, and operational management of these complex systems," said Parker with a hint of southern accent. "During NASA's Luna program in the 60s, we dominated the market with our Luna 1 rocket that put large payloads into orbit. Today we maintain some lower-value contracts with NASA, but we lost our dominance in space technology to companies like SpaceX."

"What's preventing you from gaining this dominance once again?"

"Our company's direction is driven by our management team and approved by the Board of Directors. They've taken a conservative position in the marketplace. They don't want to take risks. I don't think they realize that the space marketplace is the new frontier. If we wait too long to invest, then we won't be able to catch up with the competition," said Parker.

"So you think the Board of Directors is the main barrier to adopting a new market strategy?" Dan asked while writing notes.

"Definitely." exclaimed Parker. "We prepared an excellent cost/benefit analysis, did our homework on all areas of concern. Then two simple questions threw a wet blanket over the entire project. I just don't think they get it."

"What are the critical technology components that must work?" Dan continued.

"Wade, will you answer this question?" Parker asked. "Wade understands all of the details of our design approach and the science behind our discoveries."

"I believe Ms. Linda Cook, board member, was worried about the Deep Space Rocket," answered Wade. "And, Mr. McIntyre, whose expertise is manufacturing and supply chain management, jumped in to keep the controversy and doubt alive. Together, they can sway the other board members to vote their way. That's why Rajah Malani didn't push for a vote."

Wade continued, "Last year, our scientists were trying to create a more powerful magnet by experimenting with a combinations of rare earth metals. They discovered a magnet that would support a nuclear generator capable of running an ion thruster. The design was found to achieve thrust about seven times more powerful than a conventional rocket motor. I'm sorry to be so technical, but this thrust is enough to launch a rocket into outer space without using stages. Our team said they would be able to build a nuclear-powered ion thruster within three to five years."

"Can you tell me more about the objectives of the Aether Program, and what has changed since the board meeting?" Dan asked.

"Our objective for the Aether Program is to design a new spacecraft that can cut payload delivery cost by more than 100 times, reducing current cost from $10,000 to under $100 per payload pound. The Aether Program objective is to build and demonstrate a prototype Deep Space Rocket (DSR) motor and show NASA a spacecraft design solution that is better than the SpaceX solution. We believe NASA will then fund the building of our spacecraft," said Wade.

"What's keeping you from achieving these objectives?" Dan asked.

"We need the Board's funding approval to produce a prototype to show NASA."

Dan continued, "Therefore, AeroStar needs a $5 billion funding decision from the Board of Directors. Our engagement objective is to erase skepticism of the technology and rare-earth metal supply. Does this make sense to you at this time?"

Parker Jones said, "I believe so, but I thought you were going to do a review of our current business strategy."

"Yes, that's right," said Robert Kavanah, "We'll certainly look at your current business direction and some of the business strategy points you've documented. We'll do this to make sure that your company's actions are aligned properly and not working against each other. But as Dan implied, you have fundamental communication issues with the Board that involve highly technical concepts. If they don't understand them, then how can you get them to approve $5 billion?"

Parker and Wade looked at each other and seemed to be satisfied for the moment.

"I think this is a good engagement objective," answered Parker Jones. "We just want to get the funding to continue; otherwise, we think the company will eventually die a slow death."

Everyone paused to absorb the wording of the engagement objective.

"Before you leave, Jesse White would like to meet you and provide additional guidance."

Parker escorted them to the elevator to the executive floor where Jesse White was waiting in his office.

After introductions, Parker explained, "Robert and Dan will give us an engagement letter on Monday. I know you would like to give them some points about AeroStar and this important project." Then Parker left and closed the door.

The office was modern, with expensive paintings and an elegant statue in the corner. White's desk was completely clean except for a desk phone and his cell phone. His credenza was clean also, except for one award placard.

"What do y'all think so far?" asked Jesse White in his low southern voice.

Robert took the lead, "We're impressed with your accomplishments so far—especially the breakthroughs with your technology. We think you need us to help you get through to the board members so they will approve funding for this important initiative. This will be our focus—to help explain how and why this technology is so important to your business strategy."

"Good," said Jesse. "Our vision is clear to me. We need to succeed with changing the current space tools from old fashioned and dangerous rocket launches to ones that can land, refuel, and take off again. The space shuttle was an amazing vehicle, but it was only one step better than the heat-shielded Apollo vehicle. The shuttle cost a thousand man-hours of service for every hour of shuttle service."

"Torri Berri and Bill Anderson at ComStar are going to revolutionize space travel with their hypersonic vehicle. It's capable of taking off from an airport, flying to altitude, and then switching jet engines to our DSR motor that will launch them to high-earth orbit or deep space. Capable of managing all of its own navigation to far reaches, returning to a mid-earth orbit, and entering the atmosphere without the need for a heat shield. Its skin will be strong enough to withstand and shed enormous amounts of heat.

Once into the atmosphere, it will gradually fly to any destination on earth and land like a normal passenger airplane. Do you have any questions?"

Dan replied, "The only vehicle that could do something similar was the X-15 rocket plane built in the 1970s. It had titanium skin cooled by circulating rocket fuel on the inside. Is your design similar?"

"I can't tell you the details of our design at this time," said Jesse. "But you get the picture."

7

Santa Susana, California

WITH THE SUNRISE OVER the Los Padres National Forest in her rear view mirror, Lizzy would be in the labs early again this morning. Her commute to the test facility was short, compared to most people who live in San Fernando Valley. She loved living in Simi Valley, not only because of her short commute, but also because of her access to the best national parks in southern California. Within a one- or two-hour drive, she could be on the beautiful beaches from Santa Barbara to Malibu, or the high desert at Mojave, or the forests in Los Padres National Forest. She lived close to the most beautiful hiking and camping in all of the country.

Dr. Lizzy McKibbin received her graduate degree at Cal Tech and fell in love with southern California. She had the best job in the world at Vega's rocket test facilities. She recently made some exciting breakthroughs in nuclear physics. She had discovered a metal compound with extraordinary characteristics that would support high heat, magnetism, and neutron absorption. Her team was in the process of building the world's first nuclear generator that would power the Nelson ion thruster. The nuclear generator and the Nelson ion thruster together are called the Deep Space Rocket, or DSR. These were exciting times and she was thrilled to be a part of it. Today she was going to formulate a compound with the same formula, but she would vary the annealing times and temperatures to see if she could improve the compound characteristics.

Today she was expecting delivery of the rare earth elements she was using in the reactor core. They are pure neodymium, samarium and gadolinium from AeroStar's mining subsidiary, Annokkha Drat Exports in New Delhi, India. Her contact was Bibi Gul Shinwari, metallurgy research, whom she'd worked with for the past two years. Bibi's role was quality control for these metals, which came from mines in Afghanistan. Her lab wasn't equipped with the equipment that could validate the purity level. Bibi certified 97% purity. Gadolinium, samarium, and neodymium are shiny white metals shipped in oil to prevent reaction with oxygen and water. All three metals are used in cobalt-steel magnets. It was Lizzy's recipe that led to her breakthrough discovery.

The shipment arrived as scheduled, and she conducted her quality control of the packages to ensure no damages existed that could contaminate the samples. Then she emailed her receipt information to Bibi in New Delhi, India, signing off with "Thank you Bibi, ;-)." Since Bibi is halfway around the world, she didn't expect his reply until tomorrow.

She planned to create a new cobalt-steel magnet using the rare-earth metals. Her hypothesis for today's experiment was to improve on her original discovery by increasing density and nuetron absorption capabilities with a stronger annealing cycle. She kept the metal formula the same. Today's test was going to be a long one.

8

Park Avenue, New York

Dan was putting the finishing touches on the engagement letter when Mary Johnson arrived.

"Mr. Duggan, I'm Mary Johnson."

"Please call me Dan," he said and waved her into his office. "I'm pleased to meet you. Robert has told me a lot about you and recommended you for this engagement." he'd read Mary's CV when he attached it to the one-page engagement letter just before sending it to Robert for his review. The engagement letter simply stated the purpose of the engagement and the estimated fees, expenses, and time frame. Robert would add his polishing changes, and then print it on letterhead and sign it as Partner, Anderson & Smith LLP.

Mary was pleasant with a great disposition and presence. "Mr. Kavanah didn't tell me anything about the engagement. I'm expecting that you'll brief me. Am I correct?" she asked.

Dan looked up and noticed how she was perfectly dressed in the management consultant uniform—standard business suit with hair pulled back into a bun. Her bright smile and pretty face with little makeup, which caught him by surprise.

"Yes, ah, I'll get to that in a minute," Dan stammered. "I, I see you went to Berkeley and have a Masters from USC. Tell me why you came to Anderson & Smith."

"Well, I studied psychology because I liked the science and data analysis—understanding how people think. My first job was with IBM Global Services where I did some consulting. But I felt that I needed more human resources knowledge, so I went to USC Marshall School of Business with a concentration in Human Relations. So, I'm here because I want to work on the human relations parts of business strategy."

"That's interesting," said Dan. He was impressed with how she spoke to him, with confidence, commanding attention to her and what she said. She looked him straight in his eyes as she spoke. "This isn't a human relations engagement, so how will you contribute?"

"Yes, I know. My expertise is communicating with my subjects—I mean clients. I generally know if they are being honest with their answers. I'll help you with the client interviews."

"To complete this engagement, it will be only you and me," Dan told her. "Let me give you some background about what we know so far, and what the client is expecting us to deliver," said Dan. "NASA is outsourcing delivery of supplies to the International Space Station, and they will use outside vendors to launch extra-terrestrial vehicles for their missions to Mars. One objective they have is to lower payload delivery cost to something lower than the current $10,000 per payload pound." Dan handed Mary a copy of the engagement binder of initial research found on the web.

"Our client, Rajah Malani, CEO of AeroStar, asked us to review their business strategy. However, two board members weren't convinced the new technology would work. One had concerns about the supply chain of the rare earth metals needed to manufacture the new spacecraft. Okay so far?"

Mary nodded.

"The Board felt uncomfortable approving $5 billion given the risks. So they tabled the vote on funding until their questions are answered." He paused. "We need to explore more than just their business strategy. I understand that you've been through our 'Issue-Based Technique.' Is that correct?"

"Yes," Mary nodded. "But, I've never used it during a real engagement. Most of my work has been using 'knowledge-based' techniques from our own methodologies."

"You and I will work together to build the engagement plan. Your strategy experience will more than contribute. What is your knowledge of the aerospace industry?"

"I can handle some of the manufacturing and supply-chain terms because they cross industries. Where I need to prepare is in the space and rocketry terminology," she replied.

"I'll help you in those areas. You can start by reading up on the research I've given you in the binder. You can easily expand the searches for any particular space terminology or concept. You do not have to be an expert in space technology, but a good familiarization will help you gain confidence to conduct interviews," Dan told her. "Let's meet on Monday to work on the issues we will want to explore and create a set of hypotheses we'll want to prove. For today, why don't you read through the material I gave you and start making your own notes?"

"Dan, thank you," said Mary. "I look forward to working with you. My colleagues say I'm lucky to work for you."

"Really. What do they say?" Dan asked waiting to see how she handled the question.

"Word is, you roll up your sleeves, work long hours, and are a good mentor." She flashed him a great smile teasing him a bit. "Are you?"

"Ah, ah, yes that's a fair assessment," Dan stammered.

"I'll see you tomorrow," she replied with a big smile and left with the engagement binder. She closed his door and headed to her cubicle in the bullpen to read Dan's research papers. She smiled, thinking how good-looking he was. She'd never worked for someone with his charm and good looks before in her short career. Yes, her colleagues at IBM Global Services were all young and bright professionals, but no one had ever really struck her in the same way that Dan had. *I hope he doesn't turn out to be an egotistical jerk,* she thought to herself.

9

New Delhi, India

This MORNING HE SMILED WHEN HE OPENED his email from Dr. Lizzy with the yellow smiley face thanking him. *These Americans sure do have it easy. I wonder what she is like in person?* He hit the Reply button.

> Dr. Lizzy,
> You are welcome. Please let me know if you need more samples. How much you will need in future and when project and testing will completed?
> Your trusted advisor,
> Bibi Gul Shinwari, Metallurgy Research

HE SAT BACK AND REFLECTED on the last ten years with Annokkha Drat Exports, creating metallurgy samples of the elements they mined. He was glad to be employed, since the gas explosion on December 2, 1984, at Union Carbide India Limited (UCIL) had ended his career. Losing his stock options meant he'd be lucky to retire with a meager pension.

After he hit the Send button, he called Abdul Wazir. "Hello Abdul."

"Hello, Bibi," said Abdul. "How was last sample I sent you?"

"My American customer will need more Gadolinium soon. Will you prepare another sample for next week?" he asked.

"I may not be able to get it to you next week. I told you that I must use private courier—cash only," said Abdul. "He just raised his rates to $1,300 U.S. When can you get me the money?"

"I will get it to you in three days," replied Bibi.

Abdul hung up, walked outside his office, and dialed a number on his satellite cell phone.

10

St. Louis, Missouri

Ms. Sexton was not looking forward to the meeting with Bill and Wyatt. She had been working sixty to eighty hours a week on the spacecraft design, named Hera, the queen of heaven and the goddess of the air and constellations. Summer identified with Hera, being her creator and all. Making her able to fly through air and space was challenging.

She had spent no less than a thousand hours reviewing the X-15 design blueprints and test data from its 199 missions. The X-15 was the only successful hypersonic spacecraft to fly into outer space and land back on earth. She was amazed that it was designed and built in 1959 when digital computers were just toys for the computer scientist. The analog computer was the only real design tool available at the time.

North American Aviation employed thousands of engineers using pencil, paper, and slide rules to iterate hundreds of thousands of calculations. She studied all of the design considerations and even entered the X-15 specifications and characteristics into her computer design tool that ran on the fastest mainframe computer available from IBM. The computer can simulate many design changes to its shape, and the results of many flights and reentries to earth all in a matter of minutes. Most of her time was analyzing the results.

She looked into three other recent designs, specifically the Rockwell X-30, NASA's X-43, and the Blackswift—HTV-3X. The X-30 and X-43 were both powered by ramjets requiring massive amounts of air; they can operate only at extremely high altitudes, so she eliminated them from further review. The Blackswift design, however, matched Hera's mission, and was powered by a turbine engine for takeoff and a ramjet for high altitude. She would replace the ramjet with the DSR to launch the spacecraft into outer space.

"Good morning," said Wyatt as he and Summer entered the conference room.

"Good morning to you also," replied Bill. "I know you two are extremely busy on the Hera's design, but I thought it important to bring you up to speed on the Board's approval status." He looked into Summer's eyes to see if she'd said anything to Wyatt.

"That would be helpful to us," answered Wyatt. "But we don't have much progress to report, do we?" Wyatt also looked directly at Summer.

Summer replied, "I've made some progress, but I need to add swing-wings to improve takeoff and landing performance."

Bill said in a patronizing voice, "You know how anxious I get when progress is slow. Is there anything I may do to help? More staff, more computer equipment, anything?"

"Bill, we're getting closer each day," said Wyatt.

"Well," interrupted Bill. "The Board has given us a short rope to satisfy their needs for a funding decision. They've given us until the next Board meeting—about three months. I know you are both working long hours on this design. If you could just brief me on what's holding you up, then perhaps we can brainstorm a possible way forward. Does that make any sense?"

Looking at both of their faces, he slowly continued. "Also, they hired a consulting firm to help convince the Board to approve our funding request. They tabled additional funding votes to let the consultants answer key questions we couldn't answer at the Board meeting." He let the statement linger in the air as Wyatt and Summer absorbed the meaning.

Both Wyatt and Summer rolled their eyes and shifted in their chairs at this information, as if he'd just tied their hands behind their backs.

Wyatt jumped in with a stern voice, "Bill, how could they? We just don't have time to talk to consultants. Just let us do our jobs."

"That's what this meeting's about. What can I give the consultants that won't take much of your time, but will keep them running around gathering their data?"

Summer finally jumped into the conversation. "That seems like a diversion we don't have time for. You do realize that we're close to solving our critical design issue with Hera. Every minute I think about something else is a minute lost in solving our design problems."

"Look at it this way: if you don't solve the design problems soon, we will lose the Hera and Aether opportunities altogether," said Bill. "I believe that a little diversion will help clear your minds. Something you can surely use. Let's spend a few minutes now brainstorming ideas to keep them busy. Perhaps even some clues to solving your design problems."

"I can't imagine they would even know what questions to ask us. Especially about engineering and technical jargon," said Wyatt.

"I agree. What if I were to provide them with some of the design research you've done with the X-15 and the Blackswift?" asked Bill. "I'll give them names of the original designers to interview to learn the issues they faced back then. What do you think, Wyatt?"

"That would certainly take some of the heat off us for the time being."

"Summer, will you brief me on the top ten questions you'd need from those projects? Let's you and I meet later this week; what's a good time?"

"I'll be ready by Wednesday morning. I want you to know I'm not happy about this."

"I'm tied up all day," replied Bill. "Why don't we have an early dinner to go over the list?"

11

Green River, Wyoming

Penny's Diner on East Flaming Gorge Way was on the outskirts of town. It's a popular eating and drinking watering hole. At two o'clock most of the tables were empty, but in a south-wall booth two men were quietly drinking their coffee and engaged in conversation. Buck, the older man, was showing the younger man something on a map—fingers pointing and tracing areas of interest.

Buck was talking. "As you can see, my ranch is twenty sections of land about ten miles north of Rock Springs. We raise cattle on these areas here where we have plenty of water from springs and plenty of good grazing. Our ranch house and buildings are here.

"We are willing to give you a mineral lease for all 12,800 acres you own. It will be an open-ended lease until we can show the Bureau of Land Management the mineral contents. Then we'll file a mineral claim," said Mr. Koubiel.

"We are looking for only one specific mineral, and if we find it in ample quantities we'll finalize the contract with values based on levels in the assay report," said Kevin. "To get the sample for the assay, we will take core samples using our drilling rigs. And, when we extract the minerals, the process is done using drilling rigs. No strip mining will be used. If you agree, then we will give you a $100,000 advance on a contract that could possibly net you about $2.2 million over ten years. Does this answer your original questions?"

Kevin nodded his head as he listened to Buck. "In what locations will you do the core drilling? I would want to make sure that you don't drill near my springs—losing one or more of them would ruin our cattle operation. I would want final approval on all core locations."

"We will accommodate that request."

"Then let's go to the next step. I presume you will bring me a contract to review and that I may have my attorney review it before I sign," said Buck.

"Yes, we'll have your attorney read the nondisclosure agreement. Once signed, you and your attorney may review and sign our contract."

The two men finished their coffee, and Kevin paid the check. "I'll call you within two days to arrange our next meeting. I will tell you more about my firm at our next meeting."

Buck left the diner and climbed into his truck, heading east to get on the highway heading for Rock Springs. He was grinning as he thought about the financial gain he was about to get, but what was the mineral they were looking for? What company would want the claim on his property? What about his neighbors' properties also?

When he reached the on-ramp, he looked back as he accelerated to highway speed on Route 80. He noticed a black Chevy Suburban with dark windows accelerating in his rear view mirror—not uncommon because Interstate 80 is a major highway going east and west.

Thirty minutes later he reached the road to his property just off highway 191 north of Rock Springs and made a left turn. As he drove over the cowcatcher onto his property, he made a quick double take. The same black suburban he saw in Green River was just passing his entrance on highway 191. That's odd, he thought.

Two miles later he was at his ranch house and opened his MacBook Air to research. Hmm, trona is a major mineral mined in the Green River, what is it used for?

He typed in "trona in Wyoming."
Computer screen display:

> Trona is a relatively rare sodium-rich mineral found in the United States, Africa, China, Turkey, and Mexico. Sweetwater County, Wyoming, is a major contributor to the total world production of trona, which is mined and then processed into soda ash. Soda ash is a significant economic commodity because of its applications in manufacturing glass, chemicals, paper, detergents, textiles, paper, food, and conditioning water.

Doesn't sound like trona would be the mineral this mystery company wants? Coal is also common in Rock Springs—I'm not interested, he thought. He searched FMC Corporation because they recently discovered lithium near one of their trona mines.

He then typed in "FMC Corporation/lithium/locations."
Computer screen display:
"The FMC Corporation lithium companies are located in North Carolina, Argentina, China, and India."
FMC seems to be a worldwide player in lithium.
He typed in "what is lithium."
Computer screen display:

> Lithium is abundant in the earth's crust, about the 25th most abundant element. It is found in granites and other igneous rock formations, but in only a few special formations. A newer source for lithium is hectorite clay, the only active development of which is through the Western Lithium Corporation in the United States.

Next he tried "western lithium corporation."
Computer screen display:

Western Lithium's Kings Valley Lithium Project is located in Humboldt County in northern Nevada, approximately 100 km north-northwest of Winnemucca and 40 km west-northwest of Orovada, Nevada.

Buck wrote down two names: FMC Corporation and Western Lithium Corporation. *One of these two companies will be my mystery company and they are looking for lithium!*

12

Anderson & Smith HQ, New York—Work Day 1

Don met Mary in the office conference room he reserved for the week. She was there punctually at five minutes before 8 in the morning with her laptop, engagement binder, and work files.

"Good morning Mary. Let me plug in my laptop and put up some flipcharts I prepared," said Dan working intently with hardly a glance in Mary's direction.

"I read your initial research. It's really helpful," she said.

"You'll recognize these from your Issue-Based Technique workshop. I'll start with the Issue Diagram to better understand what the client wants, so that we can work on the issues we will address in the engagement." Dan wrote the five levels of the hierarchy.

    1. Client Objective
    2. Engagement Objective
    3. Issues
    4. Hypotheses
    5. Key Questions

"Today we'll work through the Issue-Based Technique to get the key questions to prove or disprove the hypotheses we believe will answer each issue."

"I took the training class two months ago, so the concepts are fresh in my mind," said Mary.

"Great. Then I won't bore you by telling you things you already know. I'll lead the brainstorming of ideas we come up with. We'll both document the ideas on white boards. Later we'll transfer them to my spreadsheet. If I say something that puzzles you, please ask questions."

Mary nodded, thinking to herself, *I hope he doesn't just do it his way without my contribution.*

"Sometimes I get on a roll and forget to stop to use the facilities or get refreshments; just remind me when you need a break," said Dan. "I stocked up on water and soda. Coffee pot is in the corner, ok?"

Dan wrote on the white board, "Here's AeroStar's latest Business Objective. What they want to be able to do…"

> **Client Objective:** to become the leading supplier of spacecraft, rocket motors, and operational management of these complex systems.

"Parker Jones believes the Board of Directors is his main barrier to getting the $5 billion in funding. He did not understand that their cost benefit study did not answer the real questions the Board asked: Will the technology work, and do we have adequate supply chain of the rare earth metals? This is really what we need to deliver." Dan looked at his notes before continuing. "The two Board members are Linda Cook and H. M. McIntyre."

"Sounds like the Executive Vice-President and the Aether Program leaders didn't realize that their presentation missed its mark with the Board," said Mary. "I can understand the concept of their new technology rocket motor, but I wouldn't know if it would work or not. How do we validate the technology? We're not scientists. How will we know if the suppliers of rare earth metals have enough supply?"

"You're right, so let's explore what we know about the client," Dan said, writing facts about the client on the white board. "In my his six years with Anderson & Smith, I've done consulting projects helping clients define a new or changed business strategy. Nothing like this." He looked at Mary and put his teaching hat on. "Everything I learned at the Wharton School of Business was about business strategy. For this engagement, I'm putting my engineering hat on. I remember from aeronautical school about a launch called a low-energy profile. This is a launch where the spacecraft would circle the earth, increasing speed before reaching orbit velocity. Low-energy meant just that—least possibly energy consumed. This is the launch profile selected by AeroStar. Because no rocket stages are needed, it's called a single-stage-to-orbit launch." Dan stepped to the white board and drew the earth showing a line gradually increase in altitude as it spirals around several times to reach orbit. Next to it Dan drew a world with one line arcing from surface to orbit, representing a staged launch.

"But, a single-stage-to-orbit vehicle has only been studied, never built. I found three experimental hypersonic aircraft designs capable of reaching space. Maybe the designers at AeroStar have access to some of the technology from these programs. You are right; how could we know if will work." He looked at Mary as she was listening intently.

From deep thought, Mary spoke slowly, "What if we focus our efforts on the status of their work plans for the DSR motor, and the Hera spacecraft to see if they are complete? We could investigate the status and time to complete by questioning the developers. We could put together a complete timeline showing the 'estimate to complete' for all the key design elements. And, we can ask leading questions about people's views on the project's success and what's missing."

"I like this approach! We can also get input from outside specialists who worked on the SSTO designs. Ok, let's start with the engagement objective," Dan said and wrote on the white board.

> ENGAGEMENT OBJECTIVE: To answer the Board's question, "Will the spacecraft design work?"

ISSUE 1: What is needed to complete the DSR design and build a prototype?

ISSUE 2: Does the supply chain of "rare earth metals" support the production plan?

ISSUE 3: What other technical or logistical problems exist that may hinder the spacecraft design and/or manufacture?

Mary added, "We just have to understand what they've done so far, and what is still needed in their overall timeline and plan. Are the pieces feasible? Like you said, we can also talk to other industry experts to see if they think a low-energy launch is possible."

"Mary, that's excellent! We can ask these people what they think is possible, what was holding them back when they were trying to solve the design problems of their era, and what technologies have changed since then?"

Mary thought about her own experience and education and wondered if she would be able to keep up with Dan. "I'm still worried about investigating a high technology project. You saw my degree in Psychology and MBA in Human Relations. How can I use these skills working with you? I feel that I may hold you back or, worse, hurt the project."

"Mary, you will be fine. I believe you will be able to ask leading questions to get key answers. Let's get to work on the hypotheses necessary to answer each issue question. Then we'll write out key questions to structure our interview papers. We'll end up with three or four hypotheticals per issue and lots of questions to be answered to prove each hypo true or false. Remember, a hypothetical statement is a tentative conclusion. If we prove a hypo is false, then we will identify the pieces that won't work or have serious problems."

They worked rather late framing the "issue diagram" for the engagement. Robert finally peeked into the conference room to see what they were doing and the progress they were making. "Looks like you two have a great start on the issues. You'll soon be ready to plan who you need to interview. Let's break for today and continue tomorrow morning. I want to take you to my favorite eating place for a drink and some food."

Twenty minutes later they were seated in a quiet pub frequented by Robert and Dan.

"Dan and Mary, thank you for diving into this engagement with gusto! You two seem to work well together and complement each other's skills. Mary, welcome to the firm. Do you have any questions or concerns so far?" Robert always tried to get his team relax and feel free to contribute ideas or concerns.

"I was really nervous about the client's technology, but Dan has eased my fears. He's not only an expert in aerospace, but he's a good teacher. I'm sure we'll work well together."

"Dan, any thoughts?"

"Yes, Mary and I are well into the issue diagram. She was able to see how we could validate the technology concerns of the Board. This was a breakthrough in formulating our questions."

"Excellent. Sounds like you are two are on target. Mary, solving aerospace design issues can be interesting. Dan, do you remember that story you told me when you were at Lockheed?"

"Yes I do. It was my first big project when I was a junior engineer working on the L-1011 design. I worked on the rear-engine nacelle design. We kept the engine thrust line symmetrical with the fuselage. But we didn't realize how difficult it would be to keep the airflow from stalling when the plane entered high angles of attack."

Mary shifted in her seat. "What do you mean by 'stalling'?"

Dan thought for a minute. "It's when the air loses its energy and pressure. Same principle when an airplane wing stalls: the air no longer holds the wing up; hence, the airplane falls out of the sky. Does that help?"

"I think so," she said with reserve.

"The L-1011 has an 's-shape' center engine duct for moving airflow along the top of the fuselage, around a sharp turn, then into the engine. The Boeing 727 rear engine was similar. We didn't realize that the massive volume of airflow around the bend in the intake duct was influenced by small changes in angle of attack."

Mary looked at Dan mesmerized. "Wow, I never realized that so many factors enter into a design. I always thought the L-10ll looked really sleek compared to the DC-10, but I didn't realize that the human factor was so important. What did you do?"

"We actually solved the stall issue with two key design parameters. One, we created a duct that actually restricted airflow as it turned around the bend, forcing airflow to keep from separating. Then we worked with the engine manufacturer to build in automatic power reductions as the angle of attack increased. We added an 'angle of attack' indicator that also measured intake duct pressure. The indicator would automatically throttle back, keeping the engine from stalling out."

"That's really an interesting story," said Mary with her mind spinning from all the technical jargon. "But it worries me that I don't have the engineering background that you have. How can I be credible when we interview the technical people at AeroStar?"

"Mary, you have incredible knowledge and skills in business strategy, and you also know how to read people. You won't have a preconceived notion about answers you may get. This gives you an advantage in your interview process by asking 'what do you mean' or 'how do you know that' kinds of qualifying responses rather than accepting a response at face value. Also, with your psychology background, you are better equipped to read the person rather than the answer."

Robert paid the dinner check. "Well this has been quite interesting. Both of you will do quite well on this engagement. I'll see you tomorrow."

Before they separated for home, Dan said, "Mary I want you to know that I look forward to working with you. You have worked hard to understand the client's needs and I appreciate your attention. Please don't discount your abilities to contribute; please talk to me when you have questions about the engineering stuff."

"Thanks Dan. You are supportive and I'm looking forward to learning from you." She smiled when she turned to head home. She was pleased with her day. Dan is turning out to be a really good guy to work for. And, he's really smart about this aerospace stuff.

Dan walked to Grand Central Terminal following the signs to the Metro-North train going to Greenwich, Connecticut. He was happy with the first workday on the AeroStar engagement. Mary's bright and talented—at least she figured out how we could validate the AeroStar's rocket design by looking at their work plans and talking to industry experts.

13

Santa Susana, California

Lizzy assembled the iron, cobalt, nickel, carbon, and the rare earth elements in the same molecular (mole) formula she used to create the breakthrough magnetic material she labeled MG4. After smelting the metals in their high tech lab, she poured the molten metal into four ingot bars. Then she put the bars into annealing ovens to apply temperature and time that would harden and strengthen the ingots. She varied the annealing temperatures and times for three of the ingots, labeling them MG19, MG20, and MG21. The fourth ingot, MG18, would anneal to the exact time and temperatures of current MG4, her baseline case. She believed that she could improve the magnetic properties to increase the nuclear temperatures and high absorption of neutrons in the core reactor. Even a ten percent improvement in magnetic power would have significant increase in the specific impulse of the rocket motor.

She was working with other scientists who were designing the DSR. Its design is based on the Nelson ion thruster. The nuclear thermal device generates about 250 megawatts of electric power. That is enough electric energy to start and sustain the Nelson ion thruster. The thrust would produce more than 2,000 times more thrust energy than a conventional rocket motor and would provide sufficient power to speed the spacecraft to beyond the 17,500 mph needed to leave earth's gravitational pull. The magnetic core she had discovered would make the nuclear thermal device a reality.

Five days later she completed the annealing cycles and was ready to test the new ingots of magnetic cobalt-steel; she documented her results of MG18 and compared them to her original test ingot MG4.

- Electromagnetic permeability: Failed
- Neutron absorption: Fair
- Hardness-Rockwell scale: Poor

She couldn't believe the results. This was the third sample from Bibi that didn't replicate her first discovery in MG4.

She went into her secure email server and typed:

> Dear Bibi Gul Shinwari,
>
> I am hoping to conclude my testing within the next month. For some reason, the last batch of rare earth metals does not produce the same high-quality results as the first batch you sent me. If you know of any reason the qualities are different, please let me know.
>
> 1. Please tell me the sources of the last metal shipments of Nd, Sm, and Gd. Specifically the source and purity of each sample.
>
> 2. Please send me two more batches of Nd, Sm, and Gd as soon as possible, and please provide your sources and purity.
>
> Please send via fastest possible service.
>
> Regards,
>
> Lizzy McKibbin, Ph.D. Nuclear Materials, Santa Susana Test Site

She knew that Bibi would not receive her email for about twelve hours. Meanwhile, she would pour over the data and test results one more time to determine why MG18 didn't achieve the same properties as the original MG4 sample.

14

Anderson & Smith HQ, New York—Work Days 2-4

"Hi, Mary," said Dan. "Before we start developing our tentative conclusions, hypos, for each 'issue,' do you have any questions or additions to what we've done so far?"

"No questions yet. I'm ready to start."

"Great. Here's an engagement binder we'll use to document data we collect in interviews. It contains a project work plan, scripts we'll use explaining the project during introductions, a data matrix and list of interviewees, and tabs for taking notes."

She leafed through the binder. "Dan, I see you are thorough. How do you want to conduct the interviews?"

"I'll conduct the first two or three interviews so you'll see how I control the flow of questions. I always ask permission to take notes during the introductions. It helps build rapport and trust. When we're done, we'll debrief our interview notes before we go to our next interview."

"Do you want me to ask any questions or speak?"

"Yes, you will tell the client something about yourself. But for the first two interviews, write down your questions or points, and I'll look at them before we conclude the interview. If I missed something or need clarification, then I'll continue with yours or let you speak. Eventually we'll establish our own signals and body language to hand off the questioning to each other."

"Okay."

"Also, I'd like you to use your psychology and HR skills to read each person. Are they being deceptive, or holding back answers? Sometime we get a disruptive person—I have a technique I use. I know you're an expert in a number of personality tools like Myers & Briggs. So please note anything about each person that may help us understand them like:

- Personality—what clues do you see in the office, like display of family pictures; piles of papers and documents on desk or floor versus a clean and neat desk; displays of art, awards, and degrees?

- Cooperation—does this person resist giving information based on the dialogue? What clues are being sent in body language and other physical signs of openness and cooperation?

- Authority—does this person have the necessary job skills and authority to provide answers?

- Hobbies—what hobbies or sports does this person engage in?

On the third or fourth interview, I'll let you conduct the entire interview. Any questions so far?"

"No, this is great. If I need your help or clarification, I'll ask you. Is that okay?" she asked.

"Absolutely. We are a team working together," said Dan. "One time I was conducting interviews with a new junior associate. On the fourth interview, I let him take the lead. My junior associate paused a little too long and stumbled during his opening statement. The interviewee immediately jumped at the associate saying, 'I don't have the time to waste on this.' This negative reaction caused the associate to stammer even more. I had to interrupt the interviewee to help the associate and save the interview. From that point on, I started to do a little dress rehearsal with new or inexperienced team members. I want to make everyone comfortable on his or her first interview."

"Interesting. I guess you've seen a lot in your consulting experience."

"I have. You and I will conduct two-on-one interviews to gather the data. The reason for this is to ensure we don't miss a data point—two minds are better than one when remembering details. We'll carry our binders in our briefcases and use notes and notepads. I've found that holding a giant three-ring binder is cumbersome and intimidating to the interviewee. You will quickly memorize your scripts."

"Let's continue by developing the hypotheses and key questions in our issue diagram," Dan said and pointed to the white board. "Let's brainstorm what we think might be the answers needed to prove or disprove each issue. We need to answer Issue 1: What is needed to complete the DSR design and build a prototype? We know that the DSR motor is based on two new technologies: a nuclear generator and the Nelson ion thruster. Here are my ideas about what conclusions we need to prove Issue 1 true or false."

At the white board, Dan stated each hypo as he wrote them on the board,

> Hypo 1: The DSR design and prototype work plan is complete and on target.
> Hypo 2: The Nelson ion thruster has been tested.
> Hypo 3: The nuclear generator design is complete.
> Hypo 4: The design team has adequate resources to complete the DSR prototype.

Dan paused to let Mary absorb the content. "We can't tell if the technology will work. We can only determine that they have a plan and resources to at least conduct the tests. The tests will determine if it will work. Can you think of any other hypotheses we need to answer Issue 1?"

Mary thought for a minute before responding. "Don't we also need to know if and when the nuclear generator will be tested? Can we add Hypo 5? The nuclear generator prototype for testing is on target or has been tested. Remember, both the nuclear generator and the Nelson ion thruster are critical paths to success."

"Excellent. Why didn't I see that? Any other thoughts you may have?"

"Not right now; let's walk through the questions we need to ask," said Mary.

Dan pointed at Hypo 1 and said, "Mary, what would you ask?"

"I'd want to know if they actually have a work plan to do the DSR design."

And what if I asked, "Does the team have a work plan to complete the design?

"Then I'd want to know the key milestone dates on the plan to completion."

Then Dan asked,, "When will the design be completed per the work plan?"

"I'd also want to know when they would build a prototype. They'll have to build at least one for testing."

"Mary, you are on a roll!"

They continued to brainstorm and write several more questions necessary to reach conclusion for Issue 1.

"Mary, here's a question I like to ask every interviewee about the topic in discussion. It's simple, yet powerful. Sometimes we get new pieces of information that could be critical. Since Hypo 1 is about work plans, we'd ask . . ." Dan wrote on the white board:

> Is there any detail in the work plan, no
> matter how small, that if not completed,
> would drastically change the outcomes and
> completion times in the plan?

"We'll customize this question for each hypo we're investigating."

They continued working the entire day and into late afternoon writing questions to gather the information needed to cover all hypotheses. "Let's sleep on this tonight. Tomorrow we'll revisit the questions making changes we feel necessary. Then we'll determine who we need to interview to get the answers."

15

AeroStar HQ, New York

"Do y'all really think the consultants will get the right information to convince the Board?" asked Jesse White in his native southern drawl.

"I don't understand how they will be able to answer the Board's question—will the technology work? —let alone determine if we have adequate supplies of the rare earth metals," replied Rajah. "Their engagement letter said they will assist the management team in securing the Board's approval. They are assisting us. I believe that means we must steer the consultants in the right directions to get the right data. Do you have any other thoughts before we call Harold and Torri?"

"No. They're both team players and they'll follow our lead. We'll just remind them about their bonuses. That'll get them jumpin' through hoops," said Jesse in his low guttural voice. "Ah think we limit access to only Aether Program team members we select. We'll let them interview Harold Zaben about the status of the DSR design and testing program."

"I agree. But how should we answer the rare earth metals supply? We should brief Torri about the answers he can provide about our supply chain with our own mines in Afghanistan," said Rajah.

16

Lombardo's Restaurant, St. Louis, Missouri

BILL ANDERSON DROVE OVER TO Lombardo's Restaurant on Natural Bridge Road. It was one of those blistering summer days where the heat and humidity seared everything it touched. As he drove around the airport perimeter road, the heat waves coming off the airport concrete looked like vertical waves radiating straight up, leaving a mirage that looked like the runway was flooded with water. He had a short list of questions that he considered might be asked, but he wanted Summer to lead the way forward. She knew where the skeletons were hanging, and he wanted to point the consultants in other directions. With two kids in college and a failing marriage, he couldn't afford for the project to fail. His bonus and stock options were tied to the success of the Aether Program and the Hera spacecraft design.

As he waited for Summer in a corner booth, he reflected on one particular workday last year. He was a "design engineer" again, re-living what he really enjoyed best about the aerospace field. Summer and he were debating design constraints and possible solutions when his emotional state hit high point. He let the moment gain control. It's hard to explain what happened. They were drawing the formulas on a white board showing the energy balance during a Hoffmann transfer between two circular orbits.

Their eyes met, their senses tuned in to each other, and they passionately embraced. He felt guilty, but not ashamed of what happened that night in light of his own marital problems. He knew she had moved on from that night and the others that followed. But he still held deep feelings for her. He just couldn't erase those feelings and emotions that were so vivid in his mind.

Shortly after six, she arrived and the hostess directed her to the booth in the corner.

Bill stood and greeted her, "Summer, thank you for taking this time to brief me on the talking points for the consultants. Let's order some food and get right to business."

"Yes, I'm starved and have a lot of work still on my plate for tonight." She seemed comfortable meeting Bill at the restaurant.

With the water and coffee on the table and dinners ordered, she opened the discussion, "I think you are well versed on the design approach and the history of this project, so what I'll list is what I think you should disclose." She gave him the list of eleven talking points.

Bill read the list and his jaw dropped. "This is exactly what I was looking for. Summer, you are amazing."

Hera Spacecraft Design Points

1.   A single-stage-to-orbit spacecraft with payload equal to or greater than the Space Shuttle.

2.   Take off using conventional air breathing turbofan engines to altitude of 100,000 feet and Mach 1.1.

3.   Transitioning to DSR rocket motor capable of propelling spacecraft to low earth orbit and launch to deep space transport to Mars or other systems.

4.   Spacecraft design based on X-15 and X-30 technologies using titanium cooled skin for reentry to earth's atmosphere.

5.   Spacecraft is hypersonic with swing-wings that deploy for takeoff and landing configurations.

6.   Primary fuel systems that will burn a modified jet fuel for takeoff to altitude. External fuel tanks will carry enough fuel to altitude and then be jettisoned.

7.   Internal fuel tanks that will carry enough fuel for all other flight modes including orbit to deep space and return.

8.   The DSR rocket motor will also use the jet fuel for the Nelson ion thruster. The amount of jet fuel used during ion thrust mode will be miniscule compared to fuel consumed by conventional rocket motors.

9.   At the heart of the DSR rocket motor is a nuclear generator capable of powering the Nelson ion thrusters needing about 250 megawatts of electric power to start, and less to run the ion thruster. The amount of thrust is variable, like a throttle.

10.   Jet fuel from internal fuel tanks used to cool the titanium skin during earth reentry. Heated fuel will be cooled using systems powered by the nuclear generator.

11.   Hera is in final design review stages and will be tested in quarter scale model by end of year.

"Bill, I'm sure that you will be able to handle the interview with the consultants with ease. I doubt they would be able to ask the right questions anyway. You know we still have some design issues with the swing-wing add-on. I'm confident that I'll get it done. What I really don't know is when we'll be able to complete the quarter-scale model."

17

New Delhi, India

"ABDUL, I NEED ANSWERS, QUICKLY, for Dr. Lizzy in America. I'm sending her email to you," said Bibi. "I need to know where last samples of neodymium, samarium and gadolinium you sent me come from. I thought these were from our mines. Is that so? Dr. Lizzy also wants three more samples quickly."

Abdul's heart jumped into his throat. "Let me get information for you. When I see her email I reply to you. For now I get the sources. Anything else?"

"Yes, please hurry up and package the new supplies now," said Bibi.

Abdul reached for his cellphone and dialed his contact in Kandahar. Then he called Poya, his courier, to pick up the raw minerals that he would send to Bibi. Luckily he had already ordered new batches of rare earth metals Nd, Sm, and Gd from his alternate source. But he had never asked about the type or quality of the Gd ores. He carefully wrote a new email to Bibi telling him three points, 1. Source is Kandahar mine, 2. Gd is Bastnaesite Burundi ore, and 3. New samples will be sent next week.

18

Las Vegas, Nevada

Buck Jackson took a taxi from Las Vegas International Airport to his hotel. His contact, Kevin Koubiel, sent him his itinerary on United—Rock Springs to Las Vegas. Hotel accommodations were Trump International on 2000 Fashion Show Drive. The hotel looked like a gold monolith rising sixty-four floors into the sky with big TRUMP letters affixed near the top. His suite was massive with a kitchen, black granite counters, and a lovely modern living area.

Buck walked to the windows and marveled at the views of Las Vegas Boulevard with the Mirage and Wynn hotels standing out like gems among the dozens of other hotels, jam-packed together. He rolled his carry-on into the bedroom and changed his clothes.

His burner cellphone beeped with clear instructions to meet Kevin in ten minutes.

Buck walked directly to Harrah's Casino. He entered the main entrance and let his eyes adjust before navigating to the rear food court looking for Johnny Rockets hamburgers. He grabbed a small table with two chairs and waited.

After he ordered a hamburger and chocolate shake, he saw Kevin approaching from the casino floor. Buck stood up to shake

Kevin's hand, when suddenly a triple explosion rang out and a shower of warm droplets covered his face, shirt, and jeans.

Kevin's body took flight in slow motion and landed at Buck's feet. Buck ducked the shooter's aim, and then something hard landed on his head as he thought, *I'm next.*

His head felt a rush of pain, like the whine of a tornado but with the intensity of a passing freight train; his eyes saw red. Slot machines were ding-ding-dinging in the background. Screams and commands filled his head. Buck's arms and legs froze as he collapsed or ducked; he couldn't recall.

He spoke to a cop. A squad car drove him back to Trump International, but not the front door. The cop helped him to the elevator and to his suite.

Buck's mind began to clear as the Jacuzzi massaged the outer regions of his body. But his thoughts were not yet focused into his brain.

Buck knew nothing about Kevin's employer or the mineral deal, all of which evaporated with the guy's soul. He was of no help to the authorities.

19

Port St. John, Florida
Vega Group

Vega is a white main-sequence star in the constellation Lyra. At 25 light years away, it is the fifth brightest star in the earth's sky, where it shines at an apparent visual magnitude of 0.03. Vega is the primary component of a multiple-star system. It is moving through the galaxy at a speed of 24.2 km/s relative to the sun. Its projected galactic orbit carries it between 23,900 and 25,400 light years from the center of the galaxy. It will come closest to the sun 264,000 years from now when it will brighten to magnitude –1.37 from a distance of 13.2 light years from earth. Vega has no confirmed planets since July 2013, but does have a circumstellar debris disk.

The Delta flight from LaGuardia to Palm Beach International was smooth as glass and the sky the brightest blue Dan has ever seen. He and Mary enjoyed their first class flights. They completed all preparations necessary for the upcoming interviews with key people at the Vega Group. Like all engagements, they will start by interviewing senior management involved in the project and then will drill down to the key employees.

Dan merged onto I-95 north heading to Port St John, head-quarters for Vega Group and their subsidiary, FutureDyne Rocket. Dan asked, "How do you feel so far?"

"I'm excited and a little nervous about my first interview," said Mary.

"You'll relax once we get started. I know you've reviewed the scripts, so just follow along. Remember, I'll conduct the first few interviews while you take notes."

Dan outlined the three issues and explained how they would be used to answer the engagement objective—will the technology work? Mary would explain who they would interview to gather the data.

Dan's attention was directed to the GPS system as it pointed him to take the Port St. John Parkway exit off Interstate 95. Dan turned onto US-1 and then into a parking lot for the Winn Dixie grocery store. The address for FutureDyne Rocket was 6237 US-1, and he remembered to park to the left of the grocery store.

"Mary, do you have any questions before we go in? These inter-views will set us up for our real targets, who are located at their Santa Susana Test Site in California."

"No, I'm ready," said Mary.

They were soon shaking hands with Harold Zaben, President of Vega Group, and Raymond Dabler, President of FutureDyne Rocket. The offices were modern and well appointed, but didn't have the feel of the big corporate headquarters of AeroStar in New York. While Harold and Raymond wore business suits, everyone else wore business casual—Florida style.

"Thank you for meeting with us," said Dan. "You've seen our engagement letter. We plan to provide the necessary information to the Board members to help them make an informed decision about approving funding for the Aether Program. We would like to tell you a little about our approach and ask you a few questions about the status of the project. Mary Johnson is my associate, and she will tell you about what data we plan to gather. Mary."

"Thank you, Dan," she said. "Answering your Board's question, Will the spacecraft design work? is not a question or problem that has been asked and answered before. So, we will test your plans and resources to validate the timeline. We will do this by talking directly with your project management people and interviewing key engineers and designers who are working on the project. Do you have any questions?"

Harold and Raymond looked at each other, and then Raymond spoke. "We thought that you only need to interview us about the status of the DSR propulsion unit. As you know this is a highly secret design. We don't believe that you should be bothering our people at this time. And, we're not sure that you could understand or even evaluate the science behind this research and design."

"Mary, allow me," said Dan. "We do plan to interview you today about the project status, timeline, and resources for the project. But we also need to talk to the people performing the work, see their detail plans, and understand the outstanding issues they are facing. You are right; we cannot evaluate their work, and we don't plan to." Dan paused to let them absorb this.

"We have a list of staff we would like to interview, and perhaps after we talk to you today, you may suggest additional people we should also interview. Otherwise we won't be able to reach the conclusions necessary to make a recommendation to your Board."

"Let me see the interview list," asked Harold.

Dan shared the list with Harold and Raymond for the FutureDyne Rocket employees.

"Will you excuse us for a minute while we confirm with Rajah?"

On their return, Dan and Mary conducted the interviews with both Harold and Raymond present. As usual, Dan started with open-ended questions about when and how they had made the discovery that would revolutionize space travel—the new materials that would let them design a miniature nuclear generator.

Their answers were followed up with more direct questions about the project and its current status, sources of the rare earth metals, and hurdles they anticipate in the design. After about an hour of productive questioning, Dan summarized their knowledge of the nuclear generator and the proprietary Nelson ion thruster. "We need to interview Drs. McKibbin and Costini in Santa Susana. Do you have any questions?" Mary looked at me with a smile.

"I have one request. May we get a copy of the assay report from Annokkha Drat Exports mines in India and Afghanistan?" Dan asked.

"Ah, I'm not sure I understand your question," replied Harold.

"It would be the assay report approved during the last Board meeting."

"I'll have to find out the status of that request. May I provide a copy once it arrives?"

"Thank you. Mary and I appreciate your support and look forward to working with you and your employees on this important project."

By day's end, they had scheduled interviews at the Santa Susana Test Facility in California. On their drive back to West Palm Beach, they discussed their findings from the interviews.

"Tell me what your thoughts are," asked Dan.

"First of all, it was clear to me that our interview list and scope had been limited by Rajah; otherwise, they would not have called him. I think they may be hiding something important," said Mary.

"I agree. They don't want us talking to the employees. We haven't asked to talk to the people in India or Afghanistan. Let's review our questions for Dr. Lizzy and Dr. Costini to get names of her contacts and sources of the rare earth metals."

"Ok. I'll drill deeper into the metals they have already tested," added Mary.

"Tonight, will you find out what you can about their mining operations of Annokkha Drat Exports? I will look further into the Nelson ion thruster, so that I will be equipped to ask more detailed questions too," said Dan.

Back at the hotel in West Palm Beach, they agreed to meet for dinner at seven. Dan made reservations at Lynora's on Clematis Street, one of his favorite Italian restaurants. Meanwhile they went back to their rooms to rest and do their research.

Dan went directly to his laptop and accessed the internet.

He typed: Nelson ion thruster.

The computer screen displayed:

Nelson Ion Thruster for Very High Velocity.pdf

Dan opened and read the introduction section about the research paper shared by the scientist at FutureDyne Rocket. The author was Dr. Costini. He skipped to read the conclusions and wrote notes in his journal.

Meanwhile, Mary turned on soft music and lay on the couch with her eyes closed. Her mind was working a mile a minute and she needed to relax and meditate. After a few deep breaths, her mind had cleared and her tensions relieved. She returned to the task at hand—who is Annokkha Drat Exports? She would search the web to see what's out there about the company, including AeroStar's annual report, and review their SEC filings. If she's lucky, she'll find out the names of senior management in the company.

Dan and Mary agreed on Chianti wine to accompany their Italian meals. Dan could see that Mary was tired from the day's interviews at Vega Group.

"Let's toast to our first interview," said Dan as he held up his glass. "It was a little grueling don't you think?"

"Yeah," she replied. "I really hope the interviews with the employees will be better."

"It will be. This is pretty common for management to think that they can answer all the questions."

The next morning, Dan and Mary met for breakfast and to compare their research notes.

"I found out more about the Nelson ion thruster from a research report by Dr. Costini. Her conclusion led me to believe that a workable design is pretty close. Her findings point to a breakthrough in the Nelson ion thruster technology. What did you learn?" asked Dan.

"I found little about the Annokkha Drat Exports. They have a basic informational web page with no real data. A few pictures of some of their mines. When I looked at AeroStar's Annual Report and their SEC filing, there is only a vague note about some investments in Europe and India—no company names. Their SEC statement on Strategy and Strategic Imperatives had no mention of the plan to enter the interplanetary space race. I'm thinking that they don't want to show their hand," said Mary.

"Let's focus our questions and data gathering in those areas," Dan replied. "I think we may need to go to India or perhaps Afghanistan."

20

Las Vegas, Nevada

T HE RINGING OF THE PHONE woke Buck with a start. He was still in Las Vegas. As he slowly rose to answer, the visions of a bad dream brought him back to reality. "Hello," he answered slowly.

"Is this Buck Jackson? My name is Conrad French from the Las Vegas police. I'm investigating the murder of Mr. Kevin Koubiel, and I'd like to interview you now while everything is fresh in your mind."

"Uh, I guess so. I just woke up. Perhaps in an hour or so."

"How about I order coffee in your room and we can start over a good hot cup?"

"Ok, give me a few minutes to dress," said Buck.

Conrad called room service, explained who he was, and ordered a pitcher of coffee, fruit, and Danish for Mr. Jackson's room.

Conrad had already reviewed all the statements taken by the on-site police, photos, and the coroner's preliminary notes indicating cause of death and points of entry from a large caliber hand gun—probably a 45. He worked late into the night gathering background information on the victim and Buck Jackson.

No one remembered anything about the shooter, so his next stop was the casino to view the surveillance tapes. He'd already called the manager to hold them for his review. Conrad liked to view the actual tapes on their equipment before making copies.

Conrad followed room service to Buck's room. "Hello, Mr. Jackson," said Conrad. "Not only will the coffee and Danish warm you up, but I'll give you some moral support to help you out. Are you married? Do you have a pastor or medical doctor you can see when you get home?"

"Thank you, Mr. French," said Buck as they shook hands. "I have a brother and sister in Wyoming and a network of friends. My wife is deceased."

After some small talk about Wyoming versus Las Vegas and a cup of coffee, Conrad continued, "Mr. Jackson, I need your help in trying to solve this murder. Do you feel up to answering a few questions?"

"Please call me Buck," he said. "I never saw the shooter, so I don't think I'll be of any help."

"Yes, I know. Let me ask you a little more about your relationship with the victim because that will help me understand possible motives for the murder. Anything you can tell me about your relationship will be helpful no matter how insignificant it may seem to you. Is that fair?" asked Conrad.

"Yes, I'll be glad to help the best I can."

After an hour of questioning, Conrad said, "I think that will be a good start. If I have follow-up questions, do you mind if I call you in Wyoming?"

"Yes you may."

"Excellent. Let me recap what you told me. Please correct me if misunderstood. Ok?" Conrad took a few minutes to go over his list with Buck.

"Yep, that sums it up pretty good," said Buck.

"Great. Here's my card with my cellphone and email address. Feel free to call me at any time."

"What is going to happen next? Will I have to come back to Las Vegas?" he asked.

"First, I have to find a suspect. Then I have to gather enough evidence to get an arrest warrant. That may take some time," said Conrad. "This looks like a professional hit, so I may not get very far, but I'll try. If you see the black Chevy Suburban again, try to get the license number. Thank you for your cooperation."

Back at the station, Conrad took the fingerprints from the coroner and ran them through a number of databases to see what he could find. The fingerprints would show if Mr. Koubiel had been printed for any security application, military or private, and arrests. Conrad was amazed at the vast number of matches generated. The FBI maintains the largest database of more than 51 million fingerprints of criminals, suspected criminals, government employees, military, and other special employees. This will be a good first check.

Then he ran the name "Kevin Koubiel" through the National Crime Information Center (NCIC) to see what possible criminal records may exist. He also asked the coroner to run the DNA through the Combined DNA Index System, CODIS for short. This system may turn up potential hits if Koubiel's DNA had been catalogued from a crime scene or even if a close relative's DNA had been posted.

While he was waiting for the fingerprint and NCIC results, he ran a background check with a private system used by Las Vegas Police Department (LVPD). This web bases search found a lot of information about where he might live, his credit information, marriages, and even aliases.

Mr. Koubiel's driver's license was issued from New York and showed him living in Manhattan. Once Conrad received information about his residence, telephone number, and family, he'd call NYPD to have them contact next of kin directly. He often got a lot of additional information from the local police department.

By the end of the day, Conrad had a pretty good start on a dossier on Mr. Kevin Koubiel.

```
┌─────────────────────────────────────────────┐
│                                               │
│          VICTIM INFORMATION SHEET             │
│                                               │
│   Name:  Kevin J. Koubiel                     │
│   Residence:  NY, New York                    │
│   Born:  Erie, PA                             │
│   Age:  52                                    │
│   Occupation:  Freelance Sales Representative, │
│   self employed                               │
│   Clients:  tbd                               │
│   Military/Government positions:  None        │
│   Criminal activity and arrests: None, no known │
│   association with organized crime            │
│   Income Tax Returns:  Adjusted Gross Income– │
│   $237,450; Federal Taxes Paid–$63,000        │
│   Bank Accounts: Total all Accounts–$63,450   │
│   Other Assets:                               │
│   Manhattan Apartment–$2,300,000              │
│   Florida Condo–$457,000                      │
│   Autos–$96,000                               │
│   Relationships:  TBD                         │
│   Other information:                          │
│                                               │
└─────────────────────────────────────────────┘
```

At first glance, Mr. Koubiel looked like a successful business-man. But who are his clients? Are they in the mineral or mining business? Conrad would have to look deeply into the money flow from his clients. Could one of the clients be FMC Corporation? or Western Lithium Corporation? or somehow associated with either? Then he'd have to look into the "black Suburban" angle to see if what he could find there.

21

St. Louis, Missouri

Mr. William "Bill" Anderson met Dan and Mary in the lobby and showed them to a conference room. "Would you like coffee or beverage before we get started?" he asked. "I know that you've traveled straight here from Florida."

"I would love a cold bottle of water," said Mary.

"I'll take the same, thank you." Then they took their seats in his office.

"We will first interview you about the spacecraft design since you are the VP of Research and Development and ComStar's representative on the Aether Program." Dan paused and took a sip of water.

"Then we need to interview Wyatt Calvert since he's the Project Engineer on Aether, and Dr. Summer Sexton because she is the spacecraft designer," said Dan. "I understand that you've code-named the spacecraft Hera, the Greek mythological figure, queen of heaven and goddess of the air and starry constellations—very appropriate."

"Yes, we do like to name our projects, and Hera as a sky deity does fit the role of our spacecraft's mission. Sorry to disappoint you, but I'll handle all the questions you may have for both Wyatt and Summer," said Bill in a straight, monotone voice like he'd rehearsed it a dozen times.

Mary and Dan looked at each other. Mary was now up for the rebuttal. We planned for this turn of events, knowing that Rajah had deliberately tried to limit our access to key employees. Mary's an expert in reading people and she knew exactly how to turn the discussion around to meet our needs. She already knew that Bill had not been truthful with them when they called to arrange the interviews.

"So Mr. Anderson, what's changed from our original plan that you agreed to?" Mary asked and then paused to await his reply.

Bill paused, and then took a deep breath before answering. "Both Wyatt and Summer aren't available now, but they have given me a complete outline of the project plan and design details." He handed them copies of the Hera Spacecraft Design Specifications that Summer had given him.

Dan and Mary scanned the document.

"We already know most of this from the management briefings we received," said Mary. "This is not detailed enough to help us answer critical questions about the design effort and the status of the project plan. Without the detailed data we need, we will not be able to report back to the Board in a fair and impartial way. I'm sure that you and the senior management team don't want the project to get scrapped, do you?"

Mary watched Bill's reactions carefully, noting changes in his pupils and slight hand movements, wiping his hands together. He was lying. She knew he couldn't deliver the details because only Wyatt and Summer had the knowledge about the project plan and the design.

Bill shifted uneasily in his seat before looking up at Mary. "We didn't feel comfortable sharing such technical information with you. We didn't think that you would understand it."

"We sent you a detailed list of information we need, in either paper form or accessible via your computer. We know that there could be volumes of data and reports, but we do need to review them and have access to Wyatt and Summer to answer our specific questions about the data. You didn't have any questions about our request, so we can only assume now that you don't want to provide the information. Is that correct?"

After a long pause, Bill finally replied, "We're under a lot of stress to complete the design, and both Summer and Wyatt complained about having to waste time with you. So I decided to handle your questions directly."

Mary recognized that Bill's reply was truthful.

"Mr. Anderson, please understand that we are on your side. You need us to provide an honest assessment of the project to the Board so that they will continue funding. Does that make any sense?" asked Mary. "What about the information list we sent you? Do you have the reports or access to the data?"

Bill stopped and made a short call to Wyatt. When he was done, he stood up and said, "Follow me."

He escorted them through several secured doors into a work center in the project design area.

"Welcome to the world of high tech aerospace." Dan said as he looked around the large room. He was blown away at the technology available to a working team. At one end of the room was a high-definition screen that measured about fifteen by twenty feet; both sides of the room were lined with computer terminals, printers, and graphic printing machines of various sizes. A large oval conference area took most of the center of the room. Each station had pop-up screens with keyboards, headsets, and other communication devices.

"I have an engineer on the team retrieving most of the information on your list, and the rest we can access and display here in this room," said Bill. "Both Wyatt and Summer said they will be here shortly to meet you and coordinate calendars. They will also clear their calendars to accommodate your visit. I'll take you to our cafeteria and we'll start again after lunch."

Bill had vendor badges created for Dan and Mary that would give them access to the conference room and other facilities. He told them that upon leaving, they should return the badges to the security desk; the badges would be available when they returned for follow-up visits.

Wyatt and Summer joined them with Bill, and after a few awkward minutes, Mary helped cut through the ice using her best smile. "Thank you two for assisting us in our work for the Board of Directors. Bill told us how busy you are on the Hera project—and we do understand. Please note that we will never fully understand how you do your spacecraft design work, nor will we be able to evaluate your work. That's not our purpose. But, we do fully understand the value of a good project management network diagram to show all the pieces that need to be completed. Okay?" Mary looked both Wyatt and Summer in their eyes. She could see they were starting to warm up to her.

Dan continued, "Our primary focus for this discussion will be to interview each of you. I am an aerospace engineer, so I have some understanding of the design implications. Do you have any questions?"

Mary's eyes were on Summer when she glanced briefly at Bill, who seemed to acknowledge the look with slight body language. "Yes," replied Summer as she turned to look Mary in the eyes. "Will this be our only interviews with you, or will there be more?"

"It depends on what we learn during the interview session. As we learn certain facts, we may identify additional questions that will require more time. If that's the case, then we may have to return on another date. Our schedule has already been prepared, so any follow-up questions may also be done over the phone," said Mary. "Wyatt, do you have any questions?"

Wyatt looked up from his notepad, and then to Bill and Summer. "No, I believe you explained everything."

Dan interjected, "We just received some of the reports we requested and will look at on-line information while we are here in this project room. Bill has kindly supplied one of your engineers to walk us through the on-line screens. We will spend about an hour to get through the report. Then we will meet with you, Wyatt, for about an hour. After that we'll call you, Summer, to wrap up our day. Any questions?"

For the next hour they pored over the reports that included project plans, Gantt charts, and network diagrams. Dan also asked the young engineer to run a "critical path" report for the entire Aether Program using the project management software.

Ninety minutes later, Wyatt sat down with us in the big technical conference room. Dan opened the interview with an open-ended question. "Tell us about the Hera design. How did you decide upon the hypersonic design approach?"

"We knew about the ion thruster breakthrough by the Vega Group, so we decided to build a single-stage-to-orbit spacecraft. One of the engineers folded a paper airplane that looked like this and flew it across the room. So we selected a hypersonic shape," said Wyatt.

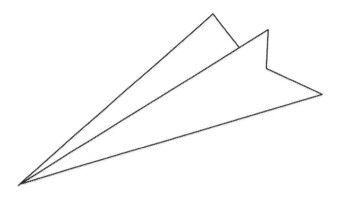

"Then we took the performance specifications for takeoff speed, payload, high earth orbit escape, reentry speeds, and landing speeds to iterate through multiple design scenarios." He stopped to take a drink of water.

"We looked at the X-15's performance specifications. We looked at Rockwell International's X-30 hypersonic space-plane. We couldn't access some of the classified design data since it was a military application, but we did research all the unclassified records," said Wyatt. "Any questions so far?"

"What design hurdles got in your way?" Dan asked.

"The hardest design part we needed to overcome was takeoff from an earth-based runway. So we added swing-wings for both takeoff and landing configurations. We're still finishing the design specs. The spacecraft will look like this when landing." Wyatt added some changes to his sketch on the tablet.

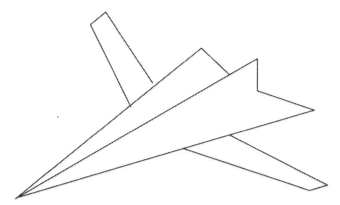

"This way, we keep the sleek hypersonic shape with wings that tuck into the sides," added Wyatt.

"How soon will you have this design configuration ready?"

"We are in the finishing touches to the configuration and will be ready for wind-tunnel testing within a week," replied Wyatt.

"So how do you carry enough fuel for a mission? It seems to me that you would need significant fuel to take off and climb to stratosphere before switching to the DSR?" Dan asked.

"What I didn't show you are the external fuel tanks that we use just for the first part of the mission. They are typical drop-tanks that will provide enough fuel for the "low-energy" climb out. The spacecraft then holds significant fuel that is used for reentry. The titanium skin is cooled by circulation over the skin of the spacecraft. Once reentry is completed, the jet turbine engines will use the remaining fuel for cruise and landing," said Wyatt.

"I understand that you spend most of your time on the master project plan that integrates the work from all the teams, including ComStar for the spacecraft and systems, Vega Group for the DSR motor and nuclear generator, and Wright Industrial for the turbofan engines. What do you believe to be the one or two critical path nodes?" asked Dan.

"Yes I spend most of my time updating and changing the project plan based on old milestones and setting of new ones. I think the critical path milestone for the entire project is the DSR motor from Vega Group. This is new technology, but we plan to build only a working prototype to show NASA that it will work," said Wyatt.

"What do you know about the DSR motor?"

"Only that our lab in California discovered a new metallurgical alloy that will support a nuclear generator that will generate more than the 250 Megawatts of power needed to run the Nelson ion thruster. The control systems for the Nelson ion thruster are much simpler than what was needed for the space shuttle. The Nelson ion thruster will be throttled with plenty of thrust to haul large payloads to deep space," said Wyatt.

After covering all their key questions with Wyatt, Dan called Summer into the project room to discuss the spacecraft design in more details.

"Thank you for your time today," said Mary. "We only have a few detail questions for you, and then we'll be on our way to California. Tell me, how long have you worked for Bill?"

"Why does that matter? I thought you wanted to know more about the spacecraft design?"

"Yes, we do, and also about the team members working on the design," replied Mary. "I get the feeling that you and Bill know a lot more, but you're not sharing. Do you mind filling in the spaces?"

"Oh, that," said Summer as she looked down, then slowly continued. "I . . . I was quite upset when told about the interviews, and I told Bill that. He agreed to intercept the interviews to let us continue working."

"Tell me more about your relationship with Bill," said Mary. She picked up a hint of something going between them by the way Summer glanced at Bill. She waited.

After what seemed like a very long pause, Summer replied in a low voice, "I've worked for Bill for about eight years in different roles. He is my mentor and he's given me more responsibility than others."

Another long pause. "On this project he picked me to be the lead design engineer and has given me a lot of good ideas from his vast experience. Unfortunately, he has problems in his home life and has tried to date me. I find it hard to work with this over my head." Another pause. "The pressure of the work is hard enough without having to deal with his feelings. I think Wyatt thinks something is going on also, but he ignores it."

"Is there anything else you can tell me about Bill that may have a bearing on the success of this project?" asked Mary.

Summer hesitated. "No, except for the marital issues at home. We're all under a lot of pressure, which may be clouding his judgment. He's been pretty indecisive the past month, and this latest interview debacle is one example."

Mary felt Summer was withholding some details, but she didn't want to lose her confidence and respect. Mary completed her interview questions but saved one more for last. "Tell me, Summer, what is the main design issue that is keeping you up at night, and if not resolved, could scuttle the entire spacecraft project?"

Summer thought a moment before carefully picking her words, "I believe it's the structural design for the swing-wing. As we scaled the size of the craft to match the payload requirements, the wing loading also increased. This was the same issue that Boeing had when they were designing the Super Sonic Transport (SST) back in the 1960s."

"What do you need to do to overcome the swing-wing issue?" asked Mary.

"I'm still working on the design and may call in some of the experts at Rockwell, as they solved the problem on the B-1 Lancer," she replied. "I could probably use some expertise in this area to shorten the design effort."

DAN AND MARY USED THE REMAINING TIME they had to summarize their notes and crosscheck to the data matrix. Mary asked, "What do you think of the swing-wing design issue? Do you think it could be a show stopper?"

"I don't know. It could certainly delay the project. I think we should bring in the lead designer at Boeing to get his input on the SST solutions," said Dan. "What do you think of Bill's leadership and his relationship with Summer?"

"There's certainly more going on than she's told us. I believe Summer has a strong personality that will overcome Bill's weaknesses. Wyatt, on the other hand, is focused on the work plan—that's good I think."

In the car, Dan asked Mary, "Do you want to have dinner together tonight?"

Mary thought a while then answered, "You know, I'm going for a jog and perhaps hit the gym for a short workout. I really need to work out the kinks in my body. I'll just order room service when I'm ready. Let's meet for breakfast tomorrow morning." Her thoughts and feelings after the dinner in New York City were fresh on her mind. Last thing she wanted was to jeopardize her job and career.

"No problem," said Dan. "Let's go over our external research plan in the morning."

22

Everett, Washington

THE NEXT MORNING, they worked on the schedule and the interview plan to meet with non-AeroStar employees. Dan and Mary heard questions being raised about the Hera variable-sweep-wing hypersonic design and wanted to confirm how other engineers from the past overcame similar design problems.

Dan set up an interview with a retired project engineer from Boeing who worked on the SST. On their flight to Seattle, Washington, Dan and Mary finalized their interview questions.

"Here are my notes on my background research of the Boeing SST and the B-1B bomber. Both had variable-sweep wings, necessary for achieving speeds greater than the speed of sound. Boeing eventually dropped out of the SST race. The hinge mechanism became so heavy that it negated the expected advantages of variable-sweep wings. The SST payload was 75,000 pounds with a cruise speed of 2.7 times the speed of sound. Rockwell's B-1B Lancer bomber flew at 1.25 times the speed of sound but carried 125,000 pounds of payload. But the Lancer's integrated-variable-sweep wing performed well with only minor adjustments," Dan said to Mary. "Remember Hera's payload is much less, at 55,250 pounds, but it must fly at hypersonic speeds in very high altitudes. I think we should find out what material technology has changed since then for the hinge mechanism." By now, Mary was well entrenched into the aerospace lingo and nodded at all the points Dan made.

"How do you want me to participate?" She asked.

"I want you to really listen, and if you hear anything that you'd like to expand on, please feel free to jump into the conversation," answered Dan. "We are meeting Mr. Lyman Masters. He retired from Boeing in 1991 and received a Distinguished Engineer title for his contribution to aircraft design where he specialized in strength of materials. He is virtually our best contact on the subject of materials and swing-wing technology. When I spoke to him, he said he spends many days on the road talking to students studying engineering. I was lucky to catch him at home, and he agreed to meet with us on such short notice."

Mr. Masters warmly welcomed Dan and Mary into his home in Everett and led them to his study. It was literally a museum of aerospace trivia including several Boeing aircraft models on stands. Dan recognized many of them by name and number: Boeing 707, Boeing 727, Boeing 747, and the Boeing SST to name a few of the early success stories. His walls were placarded with pictures of himself with past Boeing executives. He's always the center of attention—shaking hands or receiving awards. One end of the room was a full wall of books and memorabilia. And at the other end was a giant desk with piles of documents; it looked like an accountant's desk—paper binders with notes scribbled everywhere. A small laptop computer was at the center with its lid up awaiting the next command.

With a wide grin permanently affixed to his face he said, "Welcome to my humble abode. Sounds like you are trying to design another SST with a swing-wing design. How can an old engineer help you?"

He pointed to his small couch in the corner. "May I get you a beverage or snack? I have lots of time and hope I can be of help."

"Thank you for seeing us on short notice," said Dan. "We've been hired by a firm to evaluate their business strategy. We cannot tell you who our client is, but we can tell you that the strategy does include technical issues within the aerospace field."

"We would like to hear about the issues you experienced designing the SST and any specifics you can share. We did some preliminary research on the web looking at swing-wing SST and the B-1 Lancer," Dan told him. "Can you share some of the parameters you discovered? For example, what tradeoffs you made between payload and size of wing? What materials you considered for the hinge mechanism? What tradeoffs you considered and which ones you felt were best?"

Mr. Masters asked, "If you are reviewing their business strategy, then why are you getting involved with the technology? I mean, you know. I'm trying to understand the context of your question. My work fifty years ago was new in this technology."

"Our client is basing their future strategic initiatives on the success of technology similar to this. They believe it will secure a competitive advantage," said Dan. "Also, the swing-wing concept is only one piece of several technological steps they are considering."

"I see," said Lyman. "What I discovered during the SST program was that the mechanical hinge caused a significant weight problem. This was because of two situations: the hinge was too far outboard from centerline of the fuselage causing flutter and vibrations. So we had to increase the strength, which caused too much weight penalty."

"It looks to me like the Boeing wanted to keep their big delta wing shape," replied Dan.

"That's exactly right. Senior management wanted to have a look similar to the European Concord design and did not want to redesign the entire fuselage to optimally match with a swing-wing. It was a political decision that ultimately killed the SST program," he said. "They wanted to keep the sleek and long fuselage necessary for a delta wing. I told them that moving the hinge point outboard would never work. When you look at the B-1 Lancer, the wings swing from hinges at the fuselage."

"With the right configuration, would the swing-wing have worked?"

"Absolutely," exclaimed Lyman. "I couldn't convince the team to consider a radical shape that would push the speed close to hypersonic; management felt the public wasn't ready for such a big leap in technology. I believe the materials technology in today's world would support a design approaching hypersonic speeds. But I'm not sure an airline would consider such aircraft for a limited market. Virgin Galactic is building a suborbital spacecraft, but its target marketplace is the very rich."

"Mary, do you have any questions for Mr. Masters?"

"I'm impressed with your knowledge on the subject and would like to recap what we've learned so far in my own, nontechnical words."

"Yes, please recap."

"Thank you," she said. "What I learned from you is that the swing-wing should be close to the centerline for best design, and it would favor a hypersonic shape."

"Just about right," replied Lyman. "Placing the hinge near the centerline allows the hinge design to be smaller to achieve the same results."

"It's all technical to me, but it sounds to me that Boeing's team had a lot of internal issues that influenced business decisions."

"You're right, Mary," said Lyman. "But remember that the U.S. government was funding the research and they also got cold feet. When the first gas shortage hit in the mid 1970s, the price of fuel made the high-speed aircraft obsolete. All of Boeing's designs since then were slower and much more fuel efficient."

We finished our summary with Lyman Masters and headed back to our hotel in Everett, agreeing to meet in the lobby to head for dinner. "Let's consider our findings from this interview, and then discuss our plans for our meeting with Mr. Santer in Arizona," Dan said.

Back at the hotel, Dan changed into his loose warm-up pants and shirt and walked to the workout center on the first floor. In the corner of the room sat one of those old rowing machines with a sliding seat and a circular fan that increased tension based on speed of the "pull."

I love this machine. Dan strapped his feet into the stirrups, adjusted the airflow to maximum, and started his routine. He didn't see Mary enter the room to start her routine.

When Mary saw Dan working out on the rowing machine, she blushed because she'd never seen Dan in his workout clothes. He wore black and gray bike shorts that hugged tightly to his legs just above the knee and a matching shirt that revealed his muscular chest and arms. *Underneath those designer-cut business suits he has a lean and fit body. Now I know why*, she thought to herself. She went to the corner and claimed an area so she could do her stretching routine. She started in a prone position, pulling each leg to her chest. With her knees bent and her arms wide, she let her knees fall from side to side windshield-wiper style.

After a few minutes of vigorous rowing, Dan felt the heat in his thighs, quads, hamstrings, biceps, and lats. Dan slowed to end the rowing and saw Mary out of the corner of his eye doing her floor exercises. This was the first time he'd seen her in tight workout clothes.

She wore turquoise workout pants, snug to her well-shaped calves, thighs, and buttocks. Her matched top fell loosely over her hips but firm at the chest to reveal her perfectly sized breasts—not to small, not too large. She pulled her brown hair tightly to a small knot in back. She wore no makeup, her skin was clear—with a few small freckles across the bridge of her nose. Her pink lips were perfectly carved on her oval face, which revealed sparkling green eyes. Dan reached down and released the straps on his feet and swung his leg over the sliding bar. Mary was facing away from him in a prone position—Dan watched her routine in silence. From the top of her head, he could see her legs swing left to right, like windshield wipers. She was very flexible as she twisted her overlapped legs to the right, all the way to the floor with her shoulders flat while she turned her head all the way to the left.

As she held each stretch, Dan noticed her round buttocks, her narrow waistline twisted like a pretzel, and her shapely breasts unfazed by the twist but held tightly by her top.

Dan froze as he watched. Then he woke with a start when Mary said, "Hi Dan. I didn't expect to see you down here working out."

Dan blushed then stammered, "You . . . you surprised me, Mary. I didn't hear you come in." Dan's blush deepened when she rolled to her stomach and sat up on her knees. He got a glimpse of her breasts and cleavage when her top opened slightly.

"I . . . I just wanted to get a good warm-up on this rowing machine," Dan said as he felt his glowing face cool down. "How soon will you be ready for dinner?"

"I just needed to work some of the tension out of my lower back," she said. "I'll be ready in about thirty minutes. Will that work for you?"

Half an hour later, Mary came out of the elevator wearing a fashionable outfit of black slacks, loose fitting white and beige top, and black flats. She glided across the floor to meet Dan.

"Are you hungry?" asked Dan.

"Famished."

"Do you like seafood? I know a great place down by the wharf called Anthony's HomePort."

"I love it all," she replied.

Dan wasn't sure if Anthony's would be busy, so he called ahead for a reservation. He was nervous, but this was no different than the many working dinners he'd had over the past five years working at Anderson & Smith. He just has to remain objective and keep his cool.

"I'm having a glass of Chardonnay. How about yourself?" Dan asked.

"Yes, I'll have one also. As long as it's from Sonoma County."

"Ah, yes. Here's a Sonoma-Cutrer. Will it do? I didn't know that you were a wine snob."

"Yes, it's a great wine," said Mary. "I grew up in Modesto, California, but I've become fond of the Sonoma County grown wines—I think Napa wines are overpriced."

After a toast to the day's work, Dan asked Mary about her past. "I know about your education and career history. What else do you mind sharing about yourself? So far I only know you like good wines from Sonoma County."

Mary blushed a little, then smiled. "I don't know where to start. You know I'm from Modesto. It's a typical farm community made famous by Ernst and Julio Gallo. My father worked his entire career at Gallo in finance. Modesto was a really good place to grow up."

"Do you consider yourself a country girl?" he asked with a grin. "You are pretty well educated if you ask me."

"My parents both went to Berkeley. I had no choice of school," she replied. "So I thought I would follow my dream and become a psychiatrist. But the medical school requirements were too demanding."

"I really liked the math and analysis part of psychology. But when I graduated, I didn't know how I would apply it, so I went to USC for my MBA in human relations. I really feel that I'm good working with and understanding people."

"You are good. I saw how you handled hostile clients. How do you like management consulting?"

"Dan, you are good at what you do too. I'm just learning from you. You've shown me that I don't have to be an aerospace expert to gather data from the experts. That it's all about knowing what we're trying to prove or disprove."

The wine mellowed her. She relaxed. "So how did you get into this business? Tell me about the real Dan Duggan."

Dan smiled. "I started out in the aerospace field. I went to Cal Poly and got my B.S. in aerospace engineering. I wanted to be an air force pilot, fighter jets. I listened to my Dad's army stories, so I figured the military was for me."

"So why didn't you?"

"It's a long story. I had an accident parasailing along the cliffs of Torre Pines. Fractured my back. I was a damaged applicant. The Air Force wouldn't even consider me for even a commission."

"That's too bad. You don't show any signs of the injury."

"That's because I work pretty hard at keeping my back muscles strong and my joints flexible. My disc is degenerated. It causes me considerable pain all the time," said Dan. "Sometimes when I don't do my exercise routines, and when I'm under considerable job stress, my back tightens up and causes shooting pains. The only thing I can do then is lie on my back and stretch things out."

The server brought their meals and poured more wine. Mary wanted to know more about Dan's personal life, but she didn't feel comfortable asking. She knew he wasn't married, but was he seeing anyone in particular? She couldn't help but feel some butterflies flying around.

23

Tuscon, Arizona

THE AIRLINER REACHED CRUISING ALTITUDE on their flight to Tucson. Dan looked out to see the white layer of clouds hugging the coastline from Washington to California, typical for this time of year. Dan studied his notes and questions for their next interview—with Mr. Paul Santer, a former North American designer who worked on the X-15 project. Dan was glad Mr. Santer agreed to be interviewed about the X-15 design.

At one time everything about the X-15 was top secret. But nearly four decades since its last flight, he wondered if they would be able to gain some insight about its design that would help them understand the possibilities for a single-stage-to-orbit space vehicle in today's technology.

After a few minutes of thinking, he looked up just as the plane passed Mt. Shasta's eastern slope. He'd never seen Mt. Shasta in such beauty and glory. He nudged Mary on his right and nodded to the window. "Mt. Shasta."

The peak rose out of the white clouds with such beauty that it took his breath away. He knew that Mt. Whitney is the tallest peak in the lower forty-eight states at 14,505 feet. But Whitney's sharp ragged peak is difficult to spot unless you know exactly where to look. Mt. Shasta is only the fifth tallest, but probably the most beautiful of all. A nearly perfect conical shape, it is always layered with snow.

"Mary, do you know that Shasta is the home to seven named glaciers? The melting glaciers feed some of the best fisheries in the world including the upper Sacramento River. I once fished the upper Sac with my brother. The crystal clear and icy cold waters flow with such force that I had to use a balance pole to keep from falling into the rapids. My brother navigated the knee-high water flow with dexterity of an expert angler. I was really afraid of falling into the icy, rushing water."

"Nah, I'm not very good at fly fishing. But I do like many out-door sports like hiking and golfing," replied Dan. "How about you?"

"I do like the outdoors. My parents always took us camping. We went to most of the national parks. Yosemite is one of my favor-ite parks because of it beautiful waterfalls and rock formations. I really miss living near it," said Mary.

A few minutes later the pilot announced their initial descent into Tucson. Once on the ground in Tucson, they pick up a rental car and headed south on Interstate 19 to a little community called Rio Rico Norte, population 18,000. The community is just seven miles north of Nogales at the U.S./Mexico border and looks like a typical southwestern retirement neighborhood. Mr. Paul Santer's home is a modest Spanish style home with stucco siding and Spanish-style tile roofing.

Mr. Santer greeted Dan and Mary with a big warm welcome, "Bienvenido amigos. You are just in time. Please come on in." He turned and yelled, "Honey, the consultants from New York are here." Then he said, "I know you want to talk about the X-15. I'm really excited. It was my baby back in the day. It's been retired now for nearly fifty years. How can I help you?"

Paul showed them to his study at the rear of the home, facing a covered patio. His wife also introduced herself. A few minutes later she arrived with a big platter of pastries, coffee, and tea.

"Mr. Santer, thank you for meeting with us about the X-15 rocket plane," said Dan. "If you don't mind, we would like to pick your brain about some of the detail design features of the X-15. We understand that it is no longer classified and your discussion with us will not reveal secrets held by the U.S. government. Is that okay with you?"

"Yes, it is. I really don't understand how I can help you after nearly five decades. I'm not sure I remember specific details of the X-15's design," said Paul.

Dan asked a number of questions about how the design was reached. Then, specifically about how the titanium skin was cooled with circulating rocket fuel.

Paul thought about the questions for a minute, "In 1958 when we started the project, our paradigm was an airplane that flew fast. We knew nothing about hypersonic airflow and supersonic shockwave design. Our design parameters were size and shape to house the rocket motor, fuel tanks, and the pilot: hence, an airplane shape with wings, vertical and horizontal stabilizers. We actually had some help from Lockheed's skunk works with some preliminary design pictures of the SR-71. It also had titanium skin that was cooled by circulating fuel just under the skin. Our plane was air launched from a B-52, and the rocket motor took it to the speeds and altitudes to a low orbit entry."

"What was your most difficult design hurdle?" asked Dan.

"I believe it was keeping the airplane in the correct attitude during reentry from low orbit space," said Paul. "We used gyros to help keep the plane straight in space, but we ended up building the wedge-shaped vertical stabilizer that produced shockwaves at hypersonic speeds to keep the plane in line and prevent yawing. Without it, the plane could possibly enter the atmosphere sideways. That would cause a catastrophic breakup."

"What do you think of the cooled titanium skin design versus the heat shield approach used by the space shuttle?" Dan asked.

"We weren't designing a spacecraft that would be returning from deep space at more than 25,000 mph," said Paul. "The only technology available for reentry at those speeds would be a ballistic burn, where a protective surface of tiles keeps the high temperatures from destroying the spacecraft. Some of the tiles even burn away. We just had to keep the airplane oriented during the high-speed entry, but at much lower speeds than returning spacecraft."

"How would you design a hypersonic craft to reenter earth orbit, then slow down for reentry?"

"There's no escaping the need to reenter at 17,500 mph," said Paul. "The X-15 was designed only to about Mach 7; you would need your craft to enter from about Mach 25. But I'm confident that modern design, materials, and cooled skin technology could prevail. Currently, when a spacecraft returns from the moon or deep space, the first thing is to slow the spacecraft down. They orient the spacecraft backwards so the rockets face forward. A retrograde rocket-burn slows it to the necessary speed for reentry. The spacecraft must be precisely aligned at the correct angle for reentry. Too steep, the spacecraft will not survive the rapid and violent heat buildup. Too shallow, the spacecraft will skip out of the atmosphere—like a rock skipping on a pond. There is a period when the spacecraft will need to change its orientation with the nose forward so that the heat shield will deflect the extremely high temperatures." Paul paused for questions before continuing. "Now, for a sleek design like the X-30, reentry would be no problem. The cooled titanium skin will glow cherry red as the dynamic air pressure increases. It will literally fly its way into the atmosphere."

Mary said, " So you think that a modern hypersonic spacecraft could be designed to reenter earth's atmosphere from outer space? Why wasn't cooled skin technology considered by NASA before?"

"You have to understand NASA," replied Paul. "They believe in incremental steps that are safe. And once they decide on a path, they stick with it. The space shuttle was a venerable craft that performed remarkably well over its lifetime. Perhaps they will consider application of more sophisticated designs for future space travel."

We concluded our wrap-up summary and thanked Paul for his help and insights. On our drive back to the airport, Mary thought about Paul's statement of NASA's incremental approach.

"Dan, I'm having a hard time understanding NASA's 'incremental approach' that Paul mentioned. I understand they want to reduce the cost of space travel, but using Soviet reentry spacecraft and hiring SpaceX to launch vehicles seems to me like taking backward steps."

"I agree. It's like they are just 'outsourcing' rocket launches to vendors that are using the same old technology. SpaceX does have the technology to land the rocket motors on platforms and reuse them. The space shuttle did the same thing by parachuting the solid rocket motors and retrieving them in the ocean. I don't see much cost-saving leverage. The rocket launch still uses tremendous amounts of fuel to achieve the necessary launch speeds."

"It's certainly not a low-energy launch profile," Mary added.

"Mary, I think you are really starting to get this space technology lingo," he kidded her.

Their next flight was to West Palm Beach, Florida, to meet Mr. Mike Mitchell, a design engineer at Rockwell who worked on the X-30 project. After takeoff, Mary pulled out her engagement binder and went to the tab on NASA stated mission: Advanced Space Transportation Program: Paving the Highway to Space.

"Dan, NASA's goal is to reduce the cost of getting to space to hundreds of dollars per pound within 25 years. It says here quote ' . . . developing technologies that target a 100-fold reduction in the cost of getting to space by 2025, lowering the price tag to $100 per pound. As the next step beyond NASA's X-33, X-34 and X-37 flight demonstrators . . . . These advanced technologies would move space transportation closer to an airline style of operations with horizontal takeoffs and landings, quick turnaround times, and small ground support crews,'" she said. "They also say, 'an air-breathing engine—or rocket-based, combined cycle engine—gets its initial takeoff power from specially designed rockets, called air-augmented rockets, that boost performance about 15 percent over conventional rockets.'"

"Fifteen percent improvement certainly won't get a 100-fold cost reduction," Dan said. "Interesting that they mentioned air-breathing and combined cycle engine technology. It sounds similar to what AeroStar is trying to do."

"I would like to see where NASA expects to achieve their greatest cost savings," said Mary. "I think we need to ask Mr. Mike Mitchell about the X-30 capabilities and its weaknesses."

It was late when they arrived at Palm Beach International, so they took a taxi to the hotel—The Breakers Palm Beach. Dan asked the concierge to order a rental car for the drive up to Melborne in the morning. They agreed to ask Mr. Mitchell about the technology gap between the hypersonic X-30 single-stage-to-orbit spacecraft and the current SpaceX technology.

24

Melbourne, Florida

DAN WAS SEATED FOR BREAKFAST when Mary joined him. "Good morning," Dan said looking up from his paper. Dan noticed how well Mary looked in her business attire—a sleeveless white blouse with a fringe around the neck, mid-length gray skirt that couldn't help but reveal her athletic hips and legs. Mary's hair was tied back into a French braid helping to frame her perfectly oval face. *Always the perfect consultant*, he thought.

The server brought coffee and a variety of fresh squeezed juices. "Coffee is fine," Mary said. "Will you please bring me the continental plate with fresh fruit and whole wheat toast?"

"I looked at the recent X-33 project where NASA spent hundreds of millions of dollars on a craft that would never become a solution. It looked like they are doing exactly what Mr. Santer said about making incremental improvements," Mary said. "I think we need to follow up with Mr. Santer about hypersonic craft as a deep space vehicle. And to get his opinion of NASA's ability to manage successful space projects."

"I agree, but let's not get too sidetracked about NASA's strategic approach, no matter how bizarre and incongruent it seems. We need to stick to the AeroStar's strategy and look for clues as to the flaws in their plans. I want to learn more about what Mr. Mitchell thinks about a hypersonic craft that can exit and reenter earth's atmosphere. I also want to know what he thinks about using titanium skin with circulating cooling. How could this possibly work given the high reentry speeds for a craft coming from outer space?"

After driving north on I-95, they exited to Lake Washington Avenue, and, within a few minutes, turned onto Mr. Mitchell's street in Lake View Estates. The home was a large Florida style, two-story home with a large motor home in the driveway just in front of an oversized garage door with three additional doors for conventional motor vehicles.

"Welcome," said Mike Mitchell as they walked up the driveway.

"Thank you for seeing us on short notice," Dan said.

"No problem at all," he replied. "Please call me Mike. My wife and I will be heading north for a few months to visit our ten grandchildren located in three different states, so the timing is perfect. I'm not sure I can help you much since the X-30 project was canceled in 1993."

Mike showed them to the Florida room at the back of the home with large sliding glass doors facing a beautiful pool with a laminar flow waterfall quietly pouring into the deep end.

Dan went through the preliminary introductions of our engagement. Mike Mitchell would not need to reveal any top secrets that could still be active on the spacecraft's design.

That was Mary's cue to start. "I've been looking into NASA's strategic approach to future space travel that seem to be conflicting. They want to lower the cost of space travel significantly, but at the same time they have reverted back to conventional rocketry using SpaceX as a prime contractor. NASA has also spent significant monies on individual designs, like the X-30, the X-33, and others, but none seem to point them in the right direction. What is your thought about moving to a hybrid, hypersonic craft with single-stage-to-orbit or SSTO capabilities?"

"In the early 1990s, NASA, under direction of the Air Force, initiated a number of projects to investigate SSTO spacecraft. The military was looking for a military craft to deliver the star wars weapons. So their intent was not directly motivated by reducing costs," replied Mike.

"Rockwell was given the X-30 project because they had the best experience and knowledge of hypersonic aircraft. Our mission objectives were simple: 1. single-stage-to-orbit, 2. discharge payload, and 3. return to earth. The craft would then be turned around for additional missions. We had the aircraft design well configured and were working on the propulsion when the project was canceled. I did have a chance to collaborate with other designers for the X-43 and the BlackHawk-HTV-3X."

Dan jumped into the discussion, "Was there any information you can share with us that won't jeopardize your security obligations? We are interested in the design approaches you considered for the X-30. We want to see if they may be consistent considerations made by our client.

Mike thought for a minute, "I'm allowed to talk about generalities, but not the specifics of the technologies used in the X-30 design. Will that help?"

"Yes it will," said Dan. "One design parameter we'd like insight to is whether or not a hypersonic spacecraft design would be capable of returning from deep space given the much, higher speeds of reentry."

Mike said, "Reentry from deep space does present a problem with two parameters: 1. Slowing the craft before and during reentry, and 2. Keeping the craft from burning up in the atmosphere."

He continued, "I believe there are at least three technologies available now to allow these issues to be overcome, they are: 1. High power retrograde rockets to make several strategic burns to slow the craft to low earth-orbit speeds, 2. Computer and guidance system technologies that will allow for a much shallower entry and gradual descent into low earth orbit, and 3. Advanced metallurgy and skin cooling."

"The path from space is tangential to the earth moving at 25,000 miles per hour and would ricochet off the atmosphere into outer space. The craft must slow down by approaching backwards—tail first. At precise times they burn a retrograde rocket to slow down the craft," said Mike as he watched our faces. "A single burn is quite difficult to execute, so multiple burns would provide a much smoother and more precise transitioning. The Hohmann transfer formula is used to determine exactly how much energy is needed to change the orbit from, say, a relatively flat arc to an extremely sharp curve," he continued.

"Think of the best sinkerball pitcher that can throw the ball over home plate, and as the ball reaches the middle of the plate it would slow down and take a sharp turn down and hit the plate smack in the middle," he continued. "It seems physically impossible to do and would require a big force to redirect the baseball. The same is true for a large spacecraft: the G-forces are significant. That's one of the reasons for the multiple retrograde burns to slow and turn the spacecraft."

Dan said, "Mike, that's a great visual picture and analogy. How would this be done in today's spacecraft world?"

"So, today our computer technology and navigation is so good that an approach to earth could be at 26,000 miles instead of 99 miles. The velocity transfer would be less severe to the occupants with less risk of a skip out," he said.

"Do you remember Apollo 13 back in 1970? After the oxygen tank exploded on the other side of the moon, James Lovell, Jr. had to manually navigate the spacecraft. First, he had to make a rocket burn to release them from the moon's orbit, and then he had to use celestial navigation and rocket burns to make corrections to their path to earth. They had to hit an approach angle within a few degrees. You know the rest of the story."

"The Apollo program was a pure ballistic reentry—like shooting a rifle bullet into earth from outer space. They made no Hoffman transfer maneuver to turn the spacecraft."

He continued his reentry scenario, "Now in our scenario, we've achieved high-earth orbit at 26,000 miles. The spacecraft must get down to the reentry point of about 99 miles. To do so, the craft must orient itself nose first and make short burns to increase speed to get closer to earth. Think of the spacecraft as a ball on a string— the longer the string, the slower the speed of the ball. The string represents gravity forces from earth."

"Have you ever played tetherball? When you serve the ball at full length of the tether, the ball spins around the pole slowly. Then, as the tether shortens when it winds around the pole, the ball increases speed. Simple enough?" he asked.

"This is way over my head," said Mary. "But I think I can visualize a tetherball, and I do remember how the speed increased when you win."

"Excellent. Should I continue?"

They both nodded.

"Once in low earth orbit, it's time to go through the most dangerous reentry stage because the craft will be slowing from about 15,000 mph, or Mach 22, to less than 3,800 mph, or Mach 5. The space shuttle did this by burning off energy using a heat shield just like the ballistic reentry of Apollo."

"Hypersonic craft will not slow down if pointed nose first, so the craft must turn around and burn retrograde rockets to slow the craft. At the same time, it will be increasing stresses from the heated air that will circulate the spacecraft from the engine to the nose. The burn actually deflects the air, reducing the amount of friction on the skin of the craft. Once the maximum thermodynamic friction is past, the spacecraft will right itself with nose-down attitude. At Mach 5, the craft flies like a hypersonic craft, and the fuel is used to keep the skin cool."

"Wow," said Mary as she sat at the edge of her chair listening to Mike explain the flight path in layman's terms. "It sounds to me like you believe this kind of spacecraft design is possible using today's technology."

"Like I said, returning a spacecraft is only half the design profile. The spacecraft we were designing, the X-30 and the Blackswift, still had some design issues that had not been resolved."

"What do you mean?" Dan asked.

"The X-30 was designed as a hypersonic craft with scramjet propulsion. A scramjet needs only a small amount of air to burn. The Blackswift was a hybrid using conventional turbofan propulsion for takeoff, climb, and landing, but they hadn't solved the propulsion needed to get to low earth orbit and higher," said Mike. "We had similar issues with the X-30."

"So what you are telling us is that you believe that designs are possible for SSTO and return to earth?" Dan asked.

"Yes," replied Mike. "Excuse me, perhaps not SSTO because we just cannot carry enough fuel to get to orbit. You see, conventional rocket motors just do not provide the thrust needed to leave earth orbit without using stages or strap-on rockets like we've seen in the past—like the Saturn 5, the Space Shuttle, and even SpaceX's Falcon rockets. We just can't carry enough fuel for SSTO."

"What kind of breakthrough do you believe is needed to achieve a successful SSTO spacecraft?" asked Mary.

"Since I retired," replied Mike, "I've followed the research and nothing looks promising. I personally believe that the breakthrough will be with some kind of nuclear power. We won't see it in our lifetime."

"Are you familiar with ion thruster technology?" asked Dan.

"Yes, I did read a white paper on some research on this technology," said Mike, "but it still requires massive electrical power to operate."

"If one could generate the electrical power to run an ion thruster, would it generate enough thrust for a successful SSTO?" Mary asked.

"Absolutely," exclaimed Mike, "but you would need about a ton of plutonium to generate that amount of power."

When they were done with the interview, they thanked Mike for his time.

"Mike, would you mind if we call you should we have a follow-up question or two?" asked Mary. "You have been most gracious with your help."

"Yes, you may. Here's my card with my cell phone number," he said. "You may never know where I'll be. By the way, if you know of a company who can generate the power necessary for your SSTO spacecraft, let me know, as I'd be willing to invest some of my pension." He laughed and walked them to the front door.

Back in the car Dan said, "Mary, I compliment your participation in interviewing our experts. You seem to fully grasp what we're doing and how we are solving our client's issues in this engagement. I've been thinking about our next set of interviews in California. I want you to take them solo. What do you think?"

Mary thought before answering. "Are you sure we should do this, because our original thinking was to get two sets of eyes and ears on each subject?"

"That's right. But I'm thinking that you would be much more successful one-on-one with the two lady Ph.Ds. I would just inhibit the good dialogue and rapport you would establish. Besides, I know nothing about the chemistry and metallurgy of the nuclear generator. I really trust that you'll do a great job. Besides, if you run into a snag, you can always call me," said Dan.

"Thanks for your confidence," said Mary. She was smiling to herself all the way back to the airport.

25

Las, Vegas, Nevada

CON, THE NAME GIVEN TO HIM by his LVPD buddies, began documenting his case. He needed to see Mr. Koubiel's financial transactions, get telephone records, and search his New York apartment. So he contacted a Paxton Schumer, NYPD detective, and asked him to tape the apartment door shut and open a case file.

"I'll send you an affidavit so that we can subpoena his bank and telephone records," said Conrad. The affidavit would be presented to a New York judge to approve the subpoena of records. When two jurisdictions are involved, both police departments open case files and assign one or more detectives to help with the local legwork and investigation. Conrad would prepare his case as he normally would in Nevada and write an affidavit to be presented to the judge in New York. Conrad's process is to write the affidavit document, have it reviewed by a secretary for grammar, and then reviewed by another detective for clarity. Then he lets it rest for a few days before putting on the final touches. He learned by experience that a good defense attorney could twist a poorly written affidavit as evidence and get the case dismissed.

Getting the apartment yellow-taped took only a phone call. But Con also wanted to search the apartment. He needed to get approval for travel to New York, Wyoming, and wherever else the case may take him. So he continued to document his file and affidavit. He looked at his lead information sheet and shook his head.

He called Paxton Schumer. "Hello, this is Conrad French of Las Vegas Police Department. I want to thank you for your effort in sealing the apartment of Kevin Koebiel in your precinct."

"Conrad, you are welcome," replied Paxton Schumer. "I appreciate your call."

Con spent the next five minutes outlining the case with Paxton. "I think the key to this case will be identifying Mr. Koubiel's client. There has to be something illegal or underhanded going on. I need to get approval for the travel to New York and Wyoming to do some investigation. I was wondering if you would submit my affidavit to the courts to get access to his telephone records and bank statement?"

"I'd be glad to," said Paxton.

"Thanks. I look forward to meeting you when I get to New York."

26

Humboldt County, Nevada

W HILE CON WAS AWAITING the bank and phone records, he decided to check into companies that mine lithium based on Mr. Jackson's hunch and research.

He typed: western lithium corporation

The computer screen displayed:

> Western Lithium's Kings Valley Lithium Project is located in Humboldt County in northern Nevada, approximately 100 km north-northwest of Winnemucca and 40 km west-northwest of Orovada, Nevada.

He clicked on another link: Western Lithium USA Corp.

The computer screen displayed:

Western Lithium USA Corporation ("Western Lithium" or "the Company") is pleased to announce their reorganization . . .

> So Western Lithium was rebranding its name to Lithium Americas Corp. He concluded that lithium must be a hot commodity, and the corporate world is positioning their companies to win financially at the global boom.

He accessed their website and printed key telephone numbers. He started with the public relations department, knowing that he wouldn't get very far, but he knew how to work his LVPD position to his advantage. It was a long and arduous set of calls, but he finally got the name of a Vice-President of Mineral Leases.

With one call to the VP, Con reached his secretary and explained who he was and asked if he could schedule a thirty-minute call with the VP.

"Let me check his calendar," replied the assistant. She knew the VP had an open slot so she stepped into his office.

"I have a Las Vegas detective on the line investigating a murder case, and he thinks the victim was working for a lithium mineral company. Do you want to talk to him now or schedule a call?"

"What's his name?"

"Conrad French."

The VP reached for his phone. "Hello, this is Mike West. How can I help you?"

Con explained the situation of the murder at a Las Vegas casino restaurant and the mysterious deal he was negotiating with a Wyoming landowner to secure lithium mineral leases.

"In my research, I noticed that the lithium industry has been consolidating, and some companies are positioning themselves to improve their 'branding' and world reach. Have you employed a freelance sales person by the name of Kevin Koubiel to negotiate mineral rights on your behalf?"

"I don't recognize that name, but I will certainly check with my regional managers. We generally use our own employees to investigate mineral leases and then negotiate the leases directly. We have standard guidelines they follow," replied Mike.

"Do you know what companies you compete with in North America, and do you know if they use outside salespeople? I'm trying to find out who Mr. Koubiel was working for, so that I can follow that lead."

Mike paused a minute while he thought through his next point, "We have noticed the increase in competition in the lithium space. For example, FMC discovered lithium on their properties in Wyoming that look very promising. Have you spoken to them?"

"Not yet, but they are on my list to call. Do you know any other names or new entrants seeking lithium?"

"Let me think. We do look into who owns mineral leases. The Bureau of Land Management is a good source for who owns what, and the U.S. Geological Survey keeps track of mineral deposits around the world. They constantly update their databases as new discoveries are documented."

"There are about fifteen global mining companies: three major U.S. corporations, three U.K. companies, and nine Canadian companies. We are a wholly owned subsidiary of Lithium Americas Corp., a Canadian company. I'm not aware of any new entrants."

"Mr. West, thank you for taking the time now to help me unravel the possible leads to solve this case. Should you discover anything about a Kevin Koubiel from New York City, please give me a call." Con gave Mike West his contact information.

Con's next calls were to FMC—the exercise began all over again. He was able to reach someone at FMC responsible for lithium mining and leases. He did not receive any positive leads that would help.

27

Santa Susana, California

MARY PULLED INTO HER HOTEL in Simi Valley, California, after a long flight from New York's JFK and a painful drive from LAX. Los Angeles traffic was slow, heavy, and ubiquitous no matter the time of day. She was exhausted.

Before leaving New York City, she and Dan carefully updated the data matrix of key questions. They also wrote the names of the people to be interviewed whom they thought would have the answers—including Dr. McKibbin and Dr. Costini, located in the Santa Susana field laboratory leased to the Vega Group by NASA.

The labs have a long history of scientific discoveries in the nuclear field. Some of the original test bunkers were no longer available because of contamination, so the Vega Group built modern rocket test bunkers with state-of-the-art safety features. One of the labs would be used to test the DSR motor. This test bunker included a safety dome and recapture of ion thrust particles required by California emissions laws. The same test bunker would be used to test both the nuclear generator and the DSR motor in full scale.

Mary arrived at the Vega Group Santa Susana Laboratory at 7:45 a.m. She told the security guard her name and company. Ten minutes later she was driving in the facility with her temporary visitor badge, looking for the building the guard described to her. The facility, while neat and clean, looked old and dated.

She couldn't help but wonder what it was like forty years ago during its heyday. Lots of people walking around, cars spilling out of the parking lot, and sounds of progress each time a rocket motor was lit up and tested. She had read about its current state and how parts of the facility were contaminated with nuclear waste—a sad chapter in both corporation and government stewardship for the important accomplishments acknowledged by this facility.

Another security guard looked at her badge and identification, and placed a call. A few minutes later Dr. Lizzy McKibbin entered the security office and greeted Mary with a bright smile and eager demeanor. She directed Mary to a conference room reserved for the day.

"Dr. Mckibbin, perhaps you can give me some recommendations about where to do some hiking? I just cannot get over how beautiful your valley, hills, and mountains are."

"Oh, please call me Lizzy," she replied

Mary could see her posture soften a little, and after a long discussion about the wonders of the area from a naturalist's perspective, the two were approaching BFF status. Mary's homework on Lizzy had paid its dividends. Mary was now able to get her attention as she entered into the purpose of her visit.

Lizzy was fully indoctrinated and sworn to secrecy about the Company's strategic initiative. She confirmed Mary's status to review confidential data beforehand. She told Mary about how her research was the technological "breakthrough" needed to enable such a bold move in the marketplace. "The past two years have been the best years of my life working with some of the industry's best and brightest people. We are extremely excited about the prospects our technology will bring to space travel," said Lizzy.

"Tell me more about how you made the discovery."

"I was working with Chloe building a prototype Nelson ion thruster she designed. She named the design "Nelson" after her grandfather, who was her inspiration while in college. We built a quarter-scale prototype to prove the technology. We then scaled the design up to a size necessary for our spacecraft and quickly concluded the amount of power necessary to start and sustain the device was massive. I had been experimenting on metallurgical requirements for a nuclear power source. I needed a powerful cobalt magnetic source that would withstand extreme temperatures and the neutron absorption required by a nuclear generator. A core that wouldn't deteriorate over time."

"What did you discover?" asked Mary.

"I was exploring different properties of a samarium-cobalt magnet by adding small amounts of gadolinium and neodymium to give it strength and resistance to neutron deterioration. I did dozens of experiments with different alloy amounts. Then I ran different annealing profiles for each. By testing the key properties, I was able to see increased performance patterns. I followed those patterns and came up with an alloy that met all of our heat, strength, and resistance requirements."

Mary asked, "Where do you stand with demonstrating a prototype? And what steps, and how much time is needed to deliver a working DSR motor?"

"I'm in the process of duplicating my original material alloy, so that I can accumulate enough materials needed for the prototype."

"That sounds like good news. How's that going, and when do you think you will have a prototype ready?"

Lizzy hesitated a brief second and Mary caught her expression. "I . . . uh . . . I'm still gathering the necessary elements of the alloy, so I don't have a time table calculated at this time. Our 'rare earth' elements come from our mining company in India."

Mary heard the deception in her voice. "I see. Tell me more."

"Well, I have the material orders already placed and I'm awaiting word from Mr. Bibi Shinwari, head of metallurgy research at Annokkha Drat Exports in New Delhi, India. He gets the raw materials from our mines in Afghanistan, then he refines them to the purity that we request."

"When do you expect to get them, and how soon will you have the desired alloy for the prototype? Have you had problems getting the 'rare earth' elements in the past? Have you investigated the reserves of the elements and the potential 'supply chain' that would be needed to support manufacturing?"

Again, Lizzy exhibited discomfort with the question and hesitated. "I should get the current order within the month." Pause.

"If you cannot answer the questions about reserves and supply chain, that's okay. We know that your management has requested an 'assay report' from your mine in India. We are really interested in your experience so far. The Board's key concerns are will the technology work? and do you have enough rare elements to meet our current and future needs?"

"I'm really certain that the technology will work. I have a prototype design that we can build and test here in our new test stand. Chloe and I have been working a complete prototype design of the nuclear generator coupled with our Nelson ion thruster. Here's a Gantt chart of our prototype testing plan."

Mary took the document and scanned the Gantt chart, noticing that the plan would take twelve months to complete. She was surprised to see they had only one design iteration. "How confident are you that you will complete the prototype testing with only one design change iteration?"

Lizzy's demeanor stiffened a little. In a firm voice she said, "We created the Nelson ion thruster prototype with only one major design change. We, Chloe and I, believe we can duplicate that success again for the combined Nelson ion thruster coupled with the nuclear generator."

Mary and Lizzy continued the interview for another hour when Mary finally asked if it would be possible to see her labs and the new rocket test chamber.

"Our only requirement is that you cannot take any pictures."

"That's fine with me," answered Mary. "Would you be able to show me some of the 'rare earth' minerals you are using?"

"I'd be glad to." Lizzy seemed to relax.

Before entering the laboratory, Mary was given eye protection glasses, a hairnet, and a lab coat for protection. In the laboratory, she was astonished at how new and modern the equipment and facilities were. Lizzy pointed to the left side of the large room and said, "This space is where I perform my material testing, and on the right side is where Chloe Costini conducts her experiments. When we need to do a destructive test or run the Nelson ion thruster, we move to the new test facility in the next building over."

"Do you mind walking me through the steps you perform—from receiving the material samples to testing hardness and strength?"

"Yes, I'd be glad to. Through the double doors on our left is where the samples are received—usually by special courier using specialized containers." They walked to the first set of lab benches that contained several serious-looking devices. "Here, I open, label, and test each sample." Lizzy pulled open two large doors of a storage cabinet along the wall. Mary could see rows and rows of containers, labeled and color-coded. "I keep small original samples from each shipment here and my annealed alloy samples on the bottom rows."

Three shelves were labeled by the names of the rare earth minerals: neodymium, Samarium and Gadolinium. Two other shelves were labeled 'Cobalt Magnet Samples.' The samples were contained in metal boxes about five inches wide and twelve inches high and twelve inches deep. Each box had color-coded labels and numbers on the outside, making the rows of boxes look like law library of legal books. All neatly aligned.

Lizzy walked down the row and pulled three boxes: neodymium, samarium and gadolinium each marked with a green label. "Here are three samples that went into one of my successful cobalt alloys." Lizzy opened the boxes and pulled out a small metal container from each; she twisted the cap off showing Mary a small, silvery cube of metal immersed in oil.

"Neodymium improves the magnetism of cobalt steel; samarium gives the alloy strength in high temperatures; and gadolinium gives resistance to corrosion as well as neutron absorption, both especially important in a nuclear generator. But when I put all three in the alloy, given the right annealing process, the results exceed all expectations necessary for nuclear generation."

Mary was really impressed. She looked down the row of metal boxes and noticed that some had red labels. "What do the red labels mean?"

Lizzy thought a minute before answering. "I had received some batches of gadolinium that didn't provide the same annealing properties as the others. I believe they were contaminated during packaging or shipping."

"Where did the samples come from?"

"These come from our own mine in Afghanistan."

"Have they solved the quality problem?"

"Yes, they have. I'm expecting another shipment any day that were processed, packaged, and shipped with the utmost attention to detail, and we believe the quality problem is resolved."

"And if it doesn't fix the problem, then what will you do?"

"Worst case scenario would be to source samples from another company. That would be the last resort, obviously."

Mary nodded and continued the tour. Lizzy showed how the cobalt-steel alloy was smelted. Once cooled, the ingots of metal were placed into a large oven that had a computer connected with cabling.

"The computer controls the annealing time and temperature. The bars are heated to moderate temperatures and held for hours, then even higher temperatures and held for up to twenty-four hours and slowly lowered. In some cases the last part of the annealing is to quickly lower the temperature by dousing the bars in cold water," said Lizzy.

"The annealing process literally changes the molecular structure and properties of the metal. An example of annealed steel is the making of spring steel by heating, and then rapidly cooling. The spring steel bends and returns to its original position and is used to make the suspension on your car. Without annealing, the spring would bend and never return."

"Wow, I never knew how complicated this could be."

"Over here we test the final, annealed ingots of cobalt steel to measure a number of properties including hardness, strength, melting temperature, neutron absorption, and a few other tests. We load all the resulting properties back into our computers and run nuclear generator simulations," said Lizzy. She clicked the mouse at the workstation, activating a screen with tables of numbers, graphs, and pie charts.

"The final tests we perform include forging the ingots into shapes. Forging is power hammering to further improve hardness and strength of the steel."

"I've heard the term 'forging,' but I never knew what it meant," replied Mary.

"Well, forging is what the blacksmith did a hundred years ago by heating a horseshoe in a carbon kiln, then hammering it to make the necessary shapes. The blacksmith trade was handed down from father to son. They didn't know why the steel got stronger; they only know that heating, pounding, and cooling improved its strength with every step."

"Thanks for the explanation," said Mary.

Lizzy and Mary had lunch in the company cafeteria rather than drive into town. They further bonded after sharing each other's passions—nuclear research and business consulting.

After lunch and back at Lizzy's office, Mary recapped the points she learned getting consensus.

"One last question: will you give me the name of the company that would provide gadolinium as an alternate supplier, as well as other suppliers you have contact with?"

Lizzy paused while she contemplated the question. "I do not know the name of the company. I will have to get the name from Bibi Gul Shinwari in India. He's my contact from Annokkha Drat Exports, and he does similar metallurgy research of samples from our mines. May I send you the information next week?"

"That will be fine," replied Mary. "Will you also ask him to provide all other possible sources? My email address and telephone number are on my card. That's all I have at this time. May I call you if I have a follow-up question?"

Lizzy nodded and said, "Yes, that's fine. If you are ready, I will take you to Chloe Costini for your next appointment."

Mary grabbed a bottle of water from the refrigerator as Lizzy escorted her to Dr. Costini's office on the far side of the laboratory.

When Mary entered Dr. Costini's office, she noticed her office had a much warmer feel than Lizzy's. Dr. Costini had pictures all over her walls. Some were professional, probably from school, and others were of what looked like friends and family. No children.

Mary made a mental note to vary her interview techniques with her.

Mary was able to establish rapport with just a few questions to get at Dr. Costini's passion for life. She was from a large family in Brooklyn, New York. The oldest of three sisters and two brothers, her grandfather had given her the passion for science. But it wasn't until she worked on her Ph.D. at MIT in Cambridge that she really engaged in the science of plasma dynamics. She was really intrigued by changes in the physical transformation of atoms and molecules when heated to plasma state. While working on her thesis, she spent two summers working at Plasmadynamics and Electric Propulsion Laboratory (PEPL) in Michigan. She is single, but has a steady boyfriend who's an engineer at the Vega Group. They both enjoy outdoor sports including snowboarding in the Sierras and kiteboarding along the beaches in California. But her research and design work has taken priority in the past two years. She and Mary were soon on a first-name basis.

During Mary's interview with Chloe, she was able to get a good feel about the current status of the project and expectations for completion. Chloe even shared her confidential project planning reports with Mary.

"What obstacle, if any, would prevent or delay the development of the DSR motor?" asked Mary.

Chloe, being the down-to-earth person she is, wrinkled her nose and thought a moment. "I would have to say the critical path to completing the DSR motor would be the nuclear generator. I've tested a quarter scale of the Nelson ion thruster design and improved the design—we're even patenting our technological breakthroughs on the Nelson ion thruster. If the NucGen, as I call it, cannot provide the electrical power I need to run the plasma process, then all bets will be off."

"What do you think the prospects for success are at this stage in the development?"

"I really cannot say because this is Lizzy's area of expertise. But when I show you our fabrication and test facility, I'll introduce you to an engineer who's building the NucGen and the Nelson ion thruster parts. He may be someone you would be interested in interviewing. His name is Zeke Kilbaggon."

After spending an exhausting afternoon with Chloe and seeing the design fab facility, Chloe took Mary by hand into Zeke Kilbaggon's office. "Zeke, I want you to meet someone who's working for the big boys in New York, Mary Johnson."

Zeke was charming and answered Mary's questions using technical jargon that sailed over her head. "Zeke, can you bring it down ten thousand feet to my level? Just explain in layman's terms."

She finally got him to open up after explaining what her role was. Not to evaluate his performance, but to pick his brain about what's working or not.

"And how would someone like you be able to determine if the stuff we are doing 'will work' or not?" he asked.

Mary grabbed his lifeline statement and carefully stroked his ego in a subtle and clever way to keep him engaged and "on the line." It was like hooking a ten-pound steelhead trout using a five-pound line—very carefully.

"I think your knowledge is invaluable. May I schedule a couple of hours with you tomorrow? I would love to learn how the materials you use are designed and fabricated into a working solution. How about first thing tomorrow at 8 a.m.?" she asked. "If your boss objects, then you may have her call Rajah Malani's office to explain."

"Excellent. I will see you tomorrow," smiled Zeke. How could he say no?

28

Santa Susana, California
Day 2

WHILE RUNNING ALONG A WOODED PATH around the park just outside Santa Susana, she felt the air, cool and dry, flow over face and through her hair. Now I know why Lizzy and Chloe enjoy working here—all the professional freedom to explore and discover new technologies while living in an area that's paradise to the outdoors person. She ran five miles, then showered, packed, and ate a light breakfast. She planned to be at the facility to meet Zeke at eight.

"Dan, I've extended my stay here in Santa Susana to interview Mr. Zeke Kilbaggon. He's a manufacturing engineer who is building the prototype DSR motor. I think he may have some insight to the materials used in the NucGen."

"I trust your judgment. How did your interviews go with McKibbin and Costini?"

"I felt that Lizzy McKibbin seemed to be holding something back when probed with revealing questions about the NucGen metals. She's under a lot of pressure to produce results—quickly. Dr. Costini was helpful also. She has only one design iteration in the work plan. She felt the Nelson ion thruster wouldn't need a second design review. Perhaps I will learn more about the climate and culture within the organization after I interview Mr. Kilbaggon."

Zeke's office was small and cluttered with stacks of engineering documents, blueprints, research papers, and three piles of periodicals and junk mail. She laughed to herself as his personality profile, exhibited by his office, gave her clues about how she would approach the interview—direct and to the point.

"Thank you for agreeing to see me on short notice," she said. She spent a few minutes explaining the purpose of the engagement, approved by the Board of Directors, to review the business strategy for the Aether Strategic Initiative.

"Because you are working on the DSR motor, you've been cleared by internal security and sworn to secrecy. I received confirmation from Parker Jones to interview you."

"No problem. How may I help you?"

"Since you've been here for more than three years, I'd like you to fill me in on your knowledge of this project and the people working on it. Everything you tell me is confidential and will not be shared with your peers. Are you okay with that?"

"Certainly. I was hired because of my experience in design and fabrication using exotic materials. I've spent ten years with Lockheed Martin working on their top-secret military projects. If I tell you about them, they would have to shoot you." Zeke's attempt at humor didn't impress Mary.

"Tell me more about your working relationships with Lizzy and Chloe—from when you first started until now."

"Chloe was really helpful when I first started. She did a great job explaining the components I was building for the test prototypes of her Nelson ion thruster. I've just recently started to work with Lizzy on the NucGen—she hates it when I use that name. She'd say, 'Now Zeke, remember this is a real breakthrough in nuclear engineering. So don't make fun of it. It's keeping you employed.' She could really be patronizing at times."

"What have you been doing for her?"

"I've worked with the design engineers to understand Nuclear Generator parts and how they will be fabricated and assembled. I have complete freedom to acquire the latest test and fab equipment and latest in CIM—computer integrated manufacturing tools. Are you familiar with CIM?" he asked.

Mary nodded. "Yes, some of my clients are high-tech manufacturers."

"Lately I've been forge-stamping parts using the cobalt-steel ingots she's made. It hasn't been going smoothly because some of the ingots were too brittle, and fine cracks are visible. When I showed her the results she got really pissed off and stomped back to her lab with the failed part. Right now I'm waiting for her to get me more steel to forge. One time she accused me of not stamping the dyes correctly. I know my job—she cannot just point fingers."

"Let me understand," replied Mary. "The cobalt-steel magnetic core can't be forged?"

"So far, Lizzy hasn't given me a sample large enough for me to practice the fabrication of the NucGen core. I have all of the design and fabrication specifications—which requires the core to be forged to increase the strength and to shape it into a circular core that will support the generation of a small nuclear reaction."

"I understand that the critical cobalt-steel alloy material hasn't been created in quantities you need to make a prototype. Is that correct?" asked Mary.

"That's right," replied Zeke.

"What other prototype components are you working on?" asked Mary.

Zeke walked Mary to another part of his lab that had a large black object with several large black wires, like a battery jumper cable, sticking out the side. "This is one of our internal electrical storage units," he said.

"What's this used for?" asked Mary.

"The spacecraft will need a rather large battery as a power source to start all the internal equipment. The power supply requirements are rather large because of the power necessary to start the NucGen. The NucGen starter will contain a huge capacitor capability to spark it to life. The spacecraft designers have given us the specifications. We have a subcontractor making a large scale lithium ion battery that will provide all the power for the spacecraft's service module," said Zeke.

"I see," replied Mary. "So you will need a large lithium ion battery to run things?"

"Not just any lithium ion battery. This one will dwarf the kilowatt output ever used in the largest electric car. We're using lithium ion because it is light in weight and is rechargeable."

"Wow, how's this project going?" Mary asked.

"We have several prototypes from our supplier that we are testing. I just don't have a NucGen to jump start."

Mary spent another hour having Zeke show her the fab-lab as he called it—where all of his latest toys have been set up to fabricate serious parts. Zeke's charm and "geekness" blended to reveal an intelligent and motivated young man. Mary was flattered from the attention he gave her.

She returned to the conference room to pack her computer and notes and made a quick getaway to Bob Hope Airport to catch her flight to LaGuardia.

29

Las Vegas, Nevada

K EVIN KOUBIEL'S BANK STATEMENTS and telephone records were in Conrad's email box when he opened his mail on Monday. The bank records included his checking and savings accounts at CitiBank and Dime Community Bank. Both banks are chartered in the state of New York. He printed the pdf files and started looking for several types of transactions including large deposits from his clients; transfers or checks written to other banks; lists of all repetitive payments to understand his business and living costs; and unusual payees.

LVPD did have good technology to help with the detective work—latest servers, computers, and document printers. But analyzing the printed transaction records from a bank or phone company required patience, time, and his three-color technique. He clipped red, green, and blue tabs to the edge of the paper to mark transactions: green for large deposits, red for transfers, and blue for unusual transactions.

He identified several deposits that appeared to be customer payments. The oldest ones from two years ago came from two different banks—ostensibly two clients. Then eighteen months ago, a steady stream of monthly ACH payments for $12,500 per month started to flow from Midland bank. The name on the ACH transaction was U.S. Lithium Mines Corporation.

"Bingo! I found you."

Then he searched all state databases looking for U.S. Lithium Mines Corporation. He found a match from the Delaware database. A Delaware corporation named AeroSpace Materials was the holding company for U.S. Lithium Mines Corporations in Nevada, Wyoming, Utah, and Colorado. The stock ownership of Aerospace Materials was owned by only two entities—a minority interest held by Annokkha Drat Exports in India, and majority interest held by a blind trust in the United States. To break into the blind trust records would require legal action and a trip to a Delaware court.

Conrad started the subpoena process to serve on the Delaware district court to access the names in the blind trust. Since this was a murder case and he had evidence directly linking the corporation to the victim, the hearing in Delaware would be straightforward.

He then called the company telephone, which was answered by a woman. "Hello, this is U.S. Lithium Mines, may I help you?"

"May I speak with Mr. Kevin Koubiel?" he asked.

"Who's calling?"

"My name is Conrad Smith. I'm doing some work for him."

"He does not work at this location."

"May I speak to your manager?"

"He doesn't work here either," she replied.

"Will you give me his name and telephone number?" asked Conrad.

"No I don't give that information out."

"Ok, will you please give me your address? Mr. Koubiel told me to mail you my invoice for services rendered."

She reluctantly gave him an address in Brooklyn, New York.

Getting the company owners' names was his next step. He called Paxton Schumer in New York. "Hello Paxton. Thanks for your help on the Koubiel case. I'm planning to be in New York next week to search Mr. Koubiel's apartment." After some schmoozing with Paxton, he felt comfortable asking for more help. "Will you get me a search warrant to serve on U.S. Lithium Mines Corporation office in Brooklyn? I'll send you the details."

"Yes, no problem. Do you have any other information at this time?"

"Yes. I'm also serving a subpoena on Delaware district court to open access of the trustee names of the company's majority share-holders. This will get us to the bottom of their business and who's running the operation," said Conrad.

30

Flight from Las Vegas to New York LaGuardia

Air travel from Los Angeles to New York City takes about ten hours door to door. Mary's least favorite part of her job was sitting on airplanes, especially in economy class, since business class was booked full on short notice. But today's journey would probably take two or three hours longer since she had to change planes in Las Vegas. Just getting to Bob Hope Airport early in the morning would be a white-knuckle event. What should take forty-three minutes to travel the twenty-nine miles actually took seventy minutes.

She took a local highway over to I-5. Heavy stop-and-go traffic took at least thirty minutes. The last ten miles on I-5 down to North Hollywood Way was another white-knuckle event with thousands of cars all traveling together bumper-to-bumper. Watching cars duck in and out of lanes, Mary was glad she could stay in the right lane to the exit. Then it was a straight shot to the airport.

Bob Hope Airport is small compared to the big international airports. The walk from the car rental return to the check-in counter is only a block or so, but dozens of people pulling roller-boards with one ear glued to a cell phone created an obstacle course beyond compare.

She carried her computer and work papers with her but decided that she would forgo the work effort on the short leg of her journey. With a cup of coffee, she just relaxed and contemplated her findings from her Santa Susana interview with Zeke Kilbaggon.

The hop to Las Vegas International was short, and Mary walked over to the United Airlines terminal to await her flight to LaGuardia. She had forty-five minutes until boarding. In the center of the circular terminal was a gambling area with lights blinking and bells dinging, indicating that money was flowing into, or out of, the machines. Most of the money flowed in, only to be collected by the casino operators.

Mary went directly to the check-in counter and asked for an upgrade to business class using her top United Airlines status. "I'm sorry but business class has checked in full," said the agent. Working on her laptop in the cheap seats would be tight indeed. *It's a good thing I have little to write up*, she thought to herself. I'll enjoy a glass of wine and watch a video.

She boarded early and stowed her briefcase just above her. Being on the outside seat, she waited with her seatbelt off for the other passengers to arrive. She looked back at the passengers coming down the aisle and noticed a tall, handsome man awkwardly ducking overhead doors. He carried a paper bag with his lunch and a rather large, "old style" leather fold over case bursting at the seams.

"Excuse me, excuse me," he said as his case bounced off seated passengers' elbows and arms. When he reached Mary's row, he looked up at the seat numbers, then at his boarding pass, then at Mary. "I think I'm in here." He pointed to the window seat.

"No problem," said Mary as she easily glided into the aisle and stood up. Mary knew the trick to unlatch the armrest with her right hand as she pulled it up with her left—allowing her to turn her knees out into the aisle to stand up.

The man bent his tall frame in and clumsily slid into the window seat, dropping his case to the floor next to the middle seat. "Do you mind if I slide this under here if no one shows up?" he asked. "I plan to do some work on the flight."

140

Mary smiled at him and chuckled because of the sight. "No, please help yourself. Are you sure you'll be able to fit into the seat with all your stuff? Do you have a computer too?"

He nodded.

No one claimed the middle seat so she slid back into her seat and buckled up.

Once the plane was at altitude, Mary retrieved her laptop and went right to her notes. The gentleman to her right had also slid his briefcase from under the middle seat and opened the buckled flap holding it together.

Out of her peripheral vision, she could see a laptop and wires, and a half-dozen file folders with large, handwritten labels written on the tabs, and an assortment of notebooks—large ones with metal spirals like the kind you had in grade school, and several 4x6-inch ones that you flip over the top to get to the writing surface. Two of them looked quite used because the pages were puffy when folded back into the original position.

She was really curious now about this man. What does he do for a living? He certainly isn't dressed like a consultant or businessman. He is too old to be a student. He is neatly dressed in slacks and shirt, but his shoes are casual and comfortable—like maybe he walks a lot in his work.

Mary ordered a glass of red wine and he ordered sparkling water with ice. "You really are working. I hope you don't have a tight deadline or something."

"No. I just have a lot of paperwork to complete. I hope to get some done on the plane rather than in my hotel," he said glancing at Mary.

He really was working because he already had his laptop open and file folders stacked on the seat between them.

He leafed through his notepads, reading each page and making notes in the margins. Then he scrolled on his laptop using the touchpad, stopped, and typed something using two fingers.

"I see you're pretty good using two-finger method." Mary was already done typing her notes.

He slowly turned and gave Mary a perplexed scowl.

When the drinks came, Mary pulled down the center tray, "Do you mind sharing this tray for our drinks?" She gave him a big smile. He was tall with short, wavy dark hair combed straight back. He wore a closely groomed mustache with a thin goatee. And his eyes were soft and compelling.

"Gladly," he said as he looked into Mary's big green eyes and did a slight double take that made Mary chuckle to herself. "I, uh, I apologize for taking all this space." He straightened the file folders in the seat, and then went back to his computer.

She didn't want to distract him, so she just sat quietly and sipped her wine. His folders were strewn about on the seat with a couple of the notepads—one half open. She casually tried to read the words on the open notepad, but she couldn't read his hand-writing while upside down. Then she gasped when she read the handwritten name on the tab of the top file folder, **Annokkha Drat Exports**, written in big, bold letters. She couldn't quite read the second file folder tab but she could make out **U.S. Lith-**.

Mary couldn't believe what she read. Who is this person? Is he following me—a private investigator, perhaps? Is that why he didn't want to talk to me? After going over a number of scenarios, she decided to watch him closely while she pretended to read her paperback novel.

Her mind was racing a mile a minute. She suddenly felt extremely vulnerable. What was she going to do for the next four hours of flight time to LaGuardia? She concluded that he wouldn't do anything drastic while on the plane. She would be careful about what she would say.

Mary pulled out her paperback novel and opened it, pretending to read. She leaned back with her head on the left side of the seat back and positioned the book slightly to the right. She could clearly see him—reading his notes and tapping on his computer—but she couldn't make out the content on the screen. After about fifteen minutes of sitting and pretending to read, she stood up. "Will you watch my stuff?" she said to him as she got up to go to the bathroom at the end of the plane. *That was a stupid thing to say.* While waiting in line she watched his every move—to see if he looked at her book or laptop. Nothing.

Back in her seat, she decided to take a different approach. She'd flirt with him when he stopped working so intently. She decided to rest her eyes, pretending to sleep with her head and body slightly turned to the right. She relaxed, just watching through her eyelids. *This man is the slowest person I've ever seen working on a computer.*

He reached for his file folders in the middle seat and he leafed through some pages in the **Annokkha Drat Exports** folder. She noted the name on the second file folder: **U.S. Lithium Mines Corp**.

He eventually changed to the next file folder, again leafing through the pages. And finally, he opened the third file folder, name unknown. When he was done, he put the folders, pads, and computer back into his leather case and folded the large flap and snapped the buckle.

He slowly slid the briefcase under the seat and retrieved his paper bag, putting it on his tray.

"You are smart to bring food on the plane," she said, again with a big smile.

"I prefer to control what I get for food," he replied. "I generally don't eat what they sell on the plane."

"Neither do I. But I didn't have time to buy anything at the terminal in Las Vegas." *A corny reply she thought.*

"By the way, I couldn't help see you were working pretty hard. What line of work are you in?"

"I'm in law enforcement," he replied as he spread his food out.

Mary's voice tightened, "Really, what do you enforce?"

"I'm a detective—a homicide detective. How about yourself? What do you do?" he asked while eating his sandwich and chips.

Mary's head was spinning. Homicide detective—*Could be a line to throw me off.*

"I'm a business strategy consultant with Anderson & Smith in New York. You must be working on a case from all the work you are doing. Are you on a case right now?"

"Yes, I'm on a murder case, but I cannot talk about the case with you. Are you traveling on business?" he said as he relaxed to enjoy the conversation.

"Yes I am. I specialize in business strategy but, like you, I'm bound by a client non-disclosure agreement."

"Well, we can't talk about business—that's a relief!" he said with a little grin.

Changing the subject, he said quite graciously, "My name is Conrad French from Las Vegas, but call me Con. And you are?"

"I'm Mary Johnson from New York City. Pleasure to meet you." She held out her hand to complete the introduction—no wedding band. "I'm on my way home and I presume you are flying to New York on business."

"Yes I am."

"Please don't take this the wrong way, but I'm home for a few days and would be honored to show you around New York," she said in her smoothest, non-threatening way. "Or at least show you a local eatery that most tourists would never find. You do have to eat, don't you?" She knew his personality type would not be too flattered with a "come-on," so she made the offer more as a professional request.

"I have a busy schedule tomorrow, but I would welcome the local dinner offer as long as I can buy," said Conrad.

"Here's my business card," said Mary. "May I get your card also? What time will you be available tomorrow, and where are you staying?"

They finalized dinner plans and exchanged business cards. Mary felt much better about the situation, but she wasn't ready to dispel the possibility that he may be stalking her—perhaps to keep tabs on her findings or for industrial espionage purposes. But why would he have a folder labeled Annokkha Drat Exports? This was still odd—she doesn't believe in coincidences.

31

Manhattan

CONRAD WORKED THE ENTIRE DAY, first at the NYPD with detective Paxton Schumer, and then searching Mr. Koubiel's apartment. He did not find any additional information like names, telephone numbers, or email addresses that he didn't have already. He even knocked on all the doors around the apartment, but didn't learn anything new or of value for his evidence collection. He was dead tired with the time change hitting both his stomach and his head.

He couldn't keep from thinking of Mary Johnson. Are all New York professional women so direct?

His sixth sense was kicking in. I'll approach my dinner date with my senses on full alert—perhaps even do a little prying myself.

When he got to his hotel, he called Mary, "Are we still on for dinner? He then grabbed a notepad and wrote down the name of the restaurant and address. "I'll meet you there in an hour. Okay. Goodbye." His energy level suddenly increased after he took a quick shower.

ABOUT AN HOUR LATER HE ARRIVED at the restaurant, he checked his watch—I'm five minutes early.

He went inside to see if she had already arrived. The place was a cozy little boutique restaurant that smelled like Italian, but he wasn't sure.

The Maître d' asked, "Sir, do you wish to be seated?"

"I'm waiting for one more. Do you have a table that is more private?"

"Yes, I do. I'll have it set while you wait for your party."

"Thank you."

A minute later Mary walked in. He smiled when she walked up. She's lovelier than I remember, he thought to himself.

Mary had let her hair down from the tight bun, letting it fall to her shoulders in light waves. She wore a striking designer style top, a shawl, and tight black slacks with black pumps—simple, but beautiful. When she reached her hand to Conrad, he looked her straight in her beautiful green eyes and took her hand.

"Thank you for taking your time to have dinner with me."

"The pleasure is mine," she replied with a warm smile. "I don't always have the pleasure of hosting a famous Las Vegas detective!"

Once settled in their quiet booth near the back of the restaurant, they shared a little small talk about the weather and Conrad ordered a bottle of white wine before they looked at the menu. Mary told him about the food and what dishes she'd tried before to help him decide on his selection.

Conrad began the conversation. "Tell me a little about business strategy consulting. I've never known anyone in your profession."

"Well, let me see. Consulting starts with a business issue that the client needs help resolving. Clients seek our assistance for a number of reasons, for example, they may not have the skills to solve the issue, or they don't have the internal resources to do the work, or they just want to avoid the internal politics resulting from the recommendations. Strategy is a specialty that covers any number of functional issues including finance, organizational structure, customer acquisition, manufacturing, distribution, or any combination. Generally a good business strategy is difficult to duplicate by competition, and therefore achieves a competitive advantage."

"Sounds technical to me. How do you do the work?"

"We use a special consulting technique that helps us determine what data to collect and then how to interpret the findings that lead us to conclusions. Our conclusions help us make recommendations for the client to implement. I'm sure it is similar to how you gather clues to solving a murder case. You gather facts and follow sources to exhaust all possibilities. Am I right?"

"Pretty much. We also use criminology techniques that help us solve a case."

"Let me test this out," said Mary as she smiled and used her hands. "I research information about our clients also. For example, I know you are a high-profile detective solving murder cases. I learned this from newspaper articles. I know where you went to school and that you were once married." She paused to see his reaction. "When you enter a crime scene, your senses are keen to notice little things that would help you determine the facts. Am I right?"

"Yes," he answered with surprise. "Am I now your client?"

"Hold on. We do similar data gathering. When we see or hear something, it becomes a piece of data. I'll give you an example, okay?"

"Okay" he replied, not knowing where she was taking him.

"When I was on the plane from Las Vegas, I saw you working on your computer and you had a number of files in the seat between us. As an investigator and detective, do you believe in coincidences?" she asked.

"No, I don't believe in coincidences. One coincidence may raise an eyebrow. But two would certainly get my attention."

"Exactly! Sitting next to you on the plane, I saw something that really caught my attention. It was a name on one of your file folder tabs—Annokkha Drat Exports. I caught the name in my peripheral vision. You will agree that the name is not a common one."

"Are you trying to find out something about that name? Like I said, everything I do is confidential because it may be directly or indirectly related to the murder case."

"Let me tell you why the name piqued my senses. My client is a U.S. corporation that owns a minority interest in Annokkha Drat Exports, a mining company from India."

Conrad paused while he absorbed this new information before choosing his words carefully. "Yes, the company you must be working for is AeroStar," he concluded.

"Right! So is there any way we can share information without compromising our confidentiality agreements?" she asked.

"I'm following leads that link people together so that I can find a killer. I think that I may be getting close to those who could possibly link me to a suspect or suspects," said Conrad.

"And I'm interviewing employees to gather data to help with our strategy engagement. Without telling you exactly what we are doing for AeroStar, Annokkha Drat Exports' products are part of our study. I also saw the second label on the file folder, U.S. Lithium Mines Corp. I do not know them as a subsidiary of Annokkha Drat Exports or AeroStar," she said.

"Sounds like you did your homework before coming to dinner," replied Conrad. "Before I continue, I need to clarify something with you."

"What do you need to clarify? I believe I've told you what I'm doing, and I don't need to know about your investigation," said Mary.

"I won't tell you names of people I'm investigating, and I don't want to know anything about your strategy work. Does that sound okay with you?" he asked.

"Don't you often consider motives for the murder? You see, I can't think of any bigger motive than money. And business is all about money. I'm afraid that we might have an overlap—money and motive," said Mary.

"Ah, you are right. Yes, I believe money could be a strong motive. But I'm still identifying the players in the companies, and they may be players that you may need to talk to also," said Conrad.

That gave Mary a brilliant idea. One that would solve both their respective needs for confidentiality.

"What if I were to hire you onto our team and have you sign a confidentiality statement?" asked Mary. "And, you would deputize me so that I can work for you. That way we could work together without compromising each of our investigations, and we can share information so that we won't have to guess or beat around the bush."

Conrad smiled as he considered the possible working relationship. "Yes, I'm in. I'll get the appropriate documentation from my department in Las Vegas."

"And I will get an NDA from our client and put you on our team."

32

Brooklyn, New York

Back in Las Vegas, Conrad had his office e-file a motion in the Delaware courts to access information about the majority ownership of U.S. Lithium Mines Corp. The motion filed introduced evidence showing connection to Mr. Kevin Koubiel's murder. This was a standard motion filed many times before and should not present a problem getting the records opened.

Once the hearing date was scheduled, Conrad booked his trip to New York, then to Delaware, and his last stop would be Green River, Wyoming. A lot of ground to cover without a suspect, but his intuition told him that someone at, or connected with, U.S. Lithium Mines Corp. would be the culprit.

He'd also completed the paperwork to "deputize" Mary Johnson. Mary signed an investigative consultant agreement that LVPD often use for third-party investigators. It protected LVPD should Mary demand payment for her work—the agreement stated all work would be "gratis" and confidential.

Conrad called Mary from his office, "Mary, thank you for completing the deputy agreement. We can now officially work together. Deputy, my first piece of business is to review the Koubiel money trail with you. But I need you to recite the Deputy Code of Conduct to codify the agreement. Please raise your right hand and repeat after me . . . "

"Conrad, isn't this a little silly?" she replied.

"Are you holding up your right hand?"

"Ah, you can't even see me. What if I cross my finger?"

"That will jinx it," he said in his most serious, dry voice ever. "Of course I'm just kidding—a cop prank. But you're the first person to ever question the procedure before I recite the code. It's one of a number of pranks we use on the newbies. It's like a ritual and rite of passage for new cops."

"Whew. I thought maybe you where going whacky on me."

"Don't you consultants have little jokes you play on each other?"

"Nah, we're way too serious."

They both lightened up. Conrad then filled Mary in on his data and findings about Mr. Koubiel's finances and telephone contacts. He told Mary his travel agenda.

"You are welcome to meet me at Mr. Koubiel's apartment when I go through it, if it fits into your schedule. Perhaps we can brainstorm possible motives and leads within U.S. Lithium and AeroStar."

"We'll see, Conrad," replied Mary not giving him too much encouragement. He seemed eager to work with me as his new deputy, she thought to herself.

Two days later, Conrad arrived in New York and searched Mr. Koubiel's apartment. Unfortunately, Mary wasn't able to join him. Perhaps best because no new clues or findings were found.

His next visit would be U.S. Lithium's headquarters located in Brooklyn, New York. He was empowered with a subpoena to search all records, both manual and automated. He rode to Brooklyn with Paxton Schumer, since this was in his jurisdiction. Paxton had a beat cop stake out the office to observe who were going in and out, and the business operation times. He didn't want to telegraph their search intentions to the person in the office.

Conrad and Paxton took their respective positions at the front door of the office. A lady had entered the door just about thirty minutes earlier. Paxton rapped on the door with the butt of his billy club. "Open up. This is the police department."

They could hear footsteps hurrying to the door. "Hello, may I help you?" replied a brown-haired woman of about fifty with a heavy Brooklyn accent.

"Yes, we have a warrant to search this facility," said Conrad as he and Paxton held up their shields.

"Oh my God. What is going on?" asked Ruth O'Hara.

"May we come in, and I'll tell you what we are looking for?" Conrad asked.

The door opened wide as he and Paxton stepped into the office. Ruth O-Hara showed them into her office and offered them the two chairs in front. The office was small and sparsely furnished with a large old-fashioned wooden desk with a side table that held a new Dell computer. A small table was to one side with some BLM maps that Conrad recognized. On the opposite wall were a number of pullout filing cabinets. Ruth's desk was uncluttered, with a telephone, desk lamp, Rolodex-style calendar, and one legal size file—the kind with blue covers and clipped at the top with two-hole fasteners. She must have been working on this file.

Once they were all seated, Conrad continued, "Here's our search warrant. We are here because of the murder of a Mr. Kevin Koubiel. Do you him?"

Ruth gasped at the statement and nodded. "How was he . . . as you say . . . killed?"

"Before we answer that question, we have a few questions for you." Conrad went through the standard questions of name, address, telephone number, employer, boss, contact numbers, length of employment, job description, duties, and a few other occupation-related questions.

"When did you last speak with Mr. Koubiel?"

"May I look at my calendar?"

Conrad nodded.

"Yes, he called me with information about a new lease he was sending me—three weeks ago Monday," she replied.

"When did you see him last?"

"I've never met him face to face," said Ruth. "We run a very tight operation. Mr. Koubiel is meeting with potential clients and getting signed mineral leases. He then sends them to me. I process them by getting them filed with the Bureau of Land Management. We work together efficiently."

"Weren't you concerned that he hadn't called?"

"Yes I was. I left messages with the company owner, but he didn't return my calls either. So I kept coming into work. I have all the leases filed and all my land maps updated. Here, take a look at this one in Wyoming."

Ruth pulled open a wide filing cabinet that contained large BLM maps. On each map were carefully drawn outlines of properties representing lithium mineral leases. Each plot was labeled with numbers that could be traced to the leases held in a standard filing cabinet. All the maps and documents were neatly organized and labeled.

"I also manage the outside payroll service that cuts my check and Mr. Koubiel's. All payments are made electronically including our rent, utilities, and supplies. We are very efficient!" she said.

"What's the name of the company owner?" asked Conrad.

"His name is William Smith. He works in New Jersey," she replied.

With Paxton's help, they searched all the file cabinets looking for anything new that would help them identify additional people that could be involved in the business, directly or indirectly. Ms. Ruth O'Hara was more than helpful and Conrad didn't believe she was in any way connected with the murder. But he was curious about the business model. When he read the 10-K, 10-Q, and 8-K documents filed with the SEC, they provided little insight about the business—only that they were in the leasing of lands containing lithium minerals within the United States. How did they make money? I'll have to get the overview of the entire business including Annokkha Drat Exports, he thought to himself.

When they were done with the search of files and the computer, Conrad copied all the records Ruth had of the bank accounts, vendors, and contacts she had in her Rolodex.

"We need to contact Mr. William Smith and anyone connected with the business. I'm sure you understand. If you contact Mr. Smith, please tell him how urgent it is for him to call me. Better yet, please do NOT call him. We will contact him directly." Conrad gave her his card.

"But what if he calls me? What do I tell him?"

"Hmmm. Why don't you take the rest of today and tomorrow off? We do not want you to perjure yourself. If he calls you at home or on your cell phone, tell him you are sick. Do not tell him we were here. Do you understand?"

"I . . . I do," she replied rather upset.

"Once we talk to Mr. Smith at the telephone number and address you provided us, I will call you and let you know how to proceed. You may continue with your normal duties or new ones provided by your management—either Mr. Smith or someone else."

Conrad looked at Paxton. "Will you have the Newark police assign someone to follow Mr. Smith for us?"

He looked back to Ruth. "If you have any questions or concerns about your well being, please call me or Mr. Schumer at his number. Do you understand?"

"I, ah, think so," replied Ruth.

Conrad had Ruth send her boss, Mr. Smith, an email saying she was going home sick for the rest of the day. She also asked him if he had more lithium mineral leases that she could work on.

Conrad then took the keyboard from Ruth and put in an automatic forwarding email address so that he could monitor all incoming emails.

"I want you to go home now and please do not accept any calls. Do you have caller ID?"

She nodded.

"Please monitor the calls and record them for me. If you see my phone number, then please answer. Do you have any further questions? Remember this is a murder investigation and we do need your assistance. You are NOT a suspect," instructed Conrad.

Paxton stepped outside letting Conrad finalize instructions with Ruth O'Hara, while he called the Newark police and asked for a detective that he knew and trusted.

"Hello Jimmy, Paxton here," he said when his contact picked up. Paxton had worked with Jimmy on an organized crime bust that took a few months working together. The FBI coordinated a multi-state team of law enforcement detectives.

"Jimmy, I'm assisting a Las Vegas homicide detective on a case that has extended into New York. It now looks like it includes New Jersey. My gut says that the Mob has some sophisticated con going down. It appears legitimate, but I think someone is pulling the strings there in Newark."

"Give me the names, phone numbers, and addresses of your perp. I'll run them into our syndicate database to see if anything pops up."

"Jimmy, no matter what you find, we will need a search warrant because of the link to this murder case," said Paxton.

"No problem; let me do the quick search and I'll call you back." Paxton gave him Conrad's name and case number, and William Smith's name, phone, and mailing address.

On their way back to Paxton's office in Brooklyn, Paxton let Conrad know what Jimmy was doing. "Once I find out who we are dealing with, we can plan our actions."

"I'm heading to Delaware tomorrow to open a blind trust to get the names of the majority shareholders. Once we get these names, it may expand our player list. I'll let you know as soon as I get the names."

33

Dover , Delaware

CONRAD DROVE TO DOVER the night before his hearing in the U.S. district court. In his folders was evidence of the relationship and connection between Mr. Kevin Koubiel and U.S. Lithium Mines Corp. Having done this on several occasions, he knew exactly how to proceed.

Conrad met the state's attorney in an office he reserved at the courthouse and reviewed his evidence showing cause and relationship between employer, U.S. Lithium Mines Corp. and Kevin Koubiel, a hired contractor. The evidence documented the murder scene; ballistics of the murder weapon, a 9-mm pistol; and statements from office employee, telephone records and bank statements showing monthly payments.

"You are quite thorough," said the state's attorney. "On cases like this I've seen requests rejected by the court when the facts don't clearly warrant the release of trustee names. This should take less than thirty minutes with the court."

The court hearing went smoothly just as the state's attorney predicted. The judge ruled in favor of the State of Nevada, and then adjourned the session to his chambers where he opened the trustee file and reviewed the documents.

He handed the documents to Conrad. He read the names of the shareholders held in trust. The court would not allow a photograph or a Xerox copy be taken. But he did allow Conrad to write down the names of the majority shareholders in the blind trust.

Conrad consolidated his notes, and then called Mary. "Mary, I have the names in the majority shareholder blind trust of U.S. Lithium Mines Corp. I recognized one name, Rajah Malani. He's also the CEO of AeroStar."

"Wow, I never would have guessed that. What other names do you have? This may change our findings and conclusions, and perhaps our recommendations."

"There was also a Mr. William Anderson from St. Louis, and a Mr. Frankie Gallo from New Jersey," said Conrad.

"Something is going on here that I need to investigate further," replied Mary. "My client contact at ComStar in St. Louis, a subsidiary of AeroStar, is a William Anderson. Sounds to me like Mr. Anderson and Mr. Malani are in collusion with Annokkha Drat Exports in India." She paused. "I don't know a Frankie Gallo. If I had to guess, I would think that Rajah, Bill Anderson, and Annokkha Drat Exports are trying to get into the lithium mining business. Let me know what you learn about Frankie Gallo."

Conrad called Schumer with the names.

"I'll run these names in our database," said Paxton. There was a pause while he composed his thoughts. "I don't have to look up Frankie Gallo. He's a mobster known as Il Capo—into all kinds of technology and telephone scams. He came up through the ranks originally as a bookie. He most likely has someone posing as William Smith. The phone number Ruth gave us does not link directly to Gallo, but we'll probably find that it does through his intricate network of phone operations."

"I'll be back in New York tonight. Do you think we can call on Mr. Gallo tomorrow?" asked Conrad.

"I'll work on expanding the search warrant to Gallo's office and the so-called U.S. Lithium Office, but in the meantime I'll call Jimmy at the Newark PD to see if we can get some surveillance on both addresses. We'll execute the warrants at the same time so that evidence doesn't get destroyed," replied Paxton.

34

St. Louis, Missouri

D<small>AN AND</small> M<small>ARY WORKED INTO THE EVENING</small> reviewing their data and findings gathered to date, and then finalized the questions needing answers from ComStar about the Aether Program. They planned to complete the missing pieces about the spacecraft design. Specifically, they wanted to know more about the spacecraft service module that supports the main cabin guidance and environmental systems, propulsion units, and the DSR. Mary saw the size of the lithium ion battery during her interview with Mr. Kilbaggon. The lithium ion battery must be capable of starting the nuclear generator.

Bill Anderson greeted them at the lavish headquarters in St Louis, Missouri. "It's good to see you again."

Dan said, "Thank you for clearing the way for this meeting. We know how busy you are working on the project. We don't expect to take too much of your time."

Bill escorted them to his huge office on the mezzanine level of the corporate offices. "You mentioned that you're almost done with the data-gathering phase of the engagement. What can I do for you and Ms. Johnson?"

"We've made excellent progress talking to key people within the AeroStar family. We are most impressed with the reception we've received and the information we've gathered."

Dan outlined the names of the top design engineers they interviewed and gave Bill a sample of the information they received. "We're not ready to give you any conclusions yet. The external research we've done helped us relate your designs to what these former engineers felt could be reasonable using today's technology. We have questions about the status of the service module design. Mary spent two days at Santa Susana test labs, so I'll let Mary tell what she's looking for."

"Dr. McKibbin told me that ComStar is designing the service module. Can you fill us in on the details for this module?"

"How much detail do you need? I've been briefed on the design parameters."

"We want to know what the power source will be, who's building it, and anything else you can share."

"The power source for the service module will be a lithium ion battery. We selected lithium ion because it's lightweight, rechargeable, and stores significant power. The batteries will charge a bank of capacitors that will deliver a charge equivalent of 25 kilowatts of power necessary to start the nuclear generator," explained Bill. "Once the spacecraft enters the stratosphere, about twenty miles up, the pilots will start the nuclear generator in preparation to launch the spacecraft to its next destination—to service a satellite or to prepare for a deep space mission. Is this the level of detail that you are looking for?"

"Yes, this is excellent. So the service module will house the lithium battery system, and I suppose the nuclear generator will recharge the lithium batteries as needed? Where are you sourcing the lithium batteries?" she asked.

"You have that part of the service module correct. We built the service module for the Luna 1 systems that went to the moon and back, so we are experienced in this area. For Luna 1, we used fuel cells powered by liquid oxygen. We have our design specifications for the lithium batteries and are in the process of sourcing out the lithium batteries to our top vendors," said Bill.

"As you know, one important question from the Board is adequate supply. Have you investigated the lithium ion supply chain? Do you see any supply-chain problems in the future for lithium?"

"No, we don't see any future problems with supply. Lithium is the 25th most abundant element on earth, so we don't believe there will be any supply issues when we go to production."

Mary sipped water to slow the pace of the interview and to carefully prepare for her next set of questions.

"Bill, are you familiar with a new company called U.S. Lithium Mines Corp.?" She watched Bill's reaction as she paused for his answer.

"Are you asking me if I know anything about lithium mining, or a lithium mining company?" he slowly asked, almost like he was carefully pronouncing each word.

Mary hesitated slightly to see if he would continue—he is pretty good at controlling his answers.

Mary replied slowly, "No, I want to know if you know anything about a specific company, here in America, named U.S. Lithium Mines Corp."

Another pauses, then a measured reply. "I may have heard of it. Why do you ask?"

Mary recognized his technique to gain control of the interview, but she carefully asked, "Because I know who the major shareholders are. Would you like to compare notes?"

"Why are you wasting your time chasing information about a mining company when you should be working for the Board on our important strategic initiative?" Bill's face turned red and his tone became loud and threatening.

Now she'd reel him in. "Bill, please calm down. I—we—really need your help with a delicate matter that has much to do with this project and also another important event that I'm willing to share with you. But you must trust in Dan and me. Will you please hear me out?"

Bill walked to his mini-bar and took a glass from the rack. He added ice and a splash of Jack Daniels. "Dan and Mary would you like one too?"

They followed Bill to the bar and filled two glasses with ice—Mary added soda water to hers, and Dan added tap water to his.

Bill took a big drink of his Jack, "Look, I may be in a big pickle here, but first please let me know what you know, and then let's work together on a plan. I have some serious problems at home and it's really piling on the stress."

Dan interjected, "Bill, how you play this information to your personal benefit within the company and at home is your business. We are not here to judge you in any way. But, we have a few ground rules that you must follow."

"What kind of rules," asked Bill?

Mary's answered, "You cannot, I mean cannot, share what we tell you with anyone. Is that understood?"

"I'm not sure," replied Bill with a frown. "I have to report to my superiors. Is that what you meant?"

"No. You cannot talk to anyone in this company, within AeroStar, within U.S. Lithium Mines Corp., within Annokkha Drat Exports, your wife, your girlfriend. No one. Do you agree?"

Bill paused thinking, "Ok, I'll agree. Tell me what information you have and how we will work together."

"We know that you own ten thousand shares of U.S. Lithium Mines Corp. We know that Annokkha Drat Exports is a minority shareholder. We presumed you knew this. Am I right?"

Bill looked down at his drink. "Yes."

"We believe that the company was formed to secure lithium mineral leases in the four states of Colorado, Utah, Wyoming, and Nevada. We think that Annokkha Drat Exports would become the operations arm to mine the minerals from the secured leased lands. Is that correct?"

"I think so, yes."

"Bill, are you familiar with leasing operations of the company and the employees?"

"No, that is being handled in New York. I don't know who the employees are," said Bill as he gazed into his drink and swirled his glass.

"Do you know a Mr. Kevin Koubiel from New York?" asked Mary. *If he's telling me the truth, then I'll tell him*, Mary thought to herself.

Another measured pause by Bill. "No, that name doesn't ring a bell."

Now it was Dan's turn. "Bill, we are not judging you or your motives for being involved with this company. Please understand. We need your help." Dan used this to raise Bill's self-esteem and gain his trust.

"How can I help you? It seems like you know a lot more than I do and are in total command of the engagement. This information about U.S. Lithium doesn't seem necessary," replied Bill.

Mary's said, "You will be contacted by a Las Vegas homicide detective by the name of Conrad French. He's investigating the murder of Mr. Kevin Koubiel. Mr. Koubiel was working for U.S. Lithium to secure the mineral leases." Mary watched Bill's face turn ashen gray.

"I didn't do it nor did I even know the man."

"We don't believe you are involved either. We didn't want the detective to barge in here with a search warrant. So we made a deal with Mr. French to talk to you first and hatch a plan to keep you quiet in the investigation. He will call you, so please cooperate with him—he's thorough. Do you have any other information about U.S. Lithium Mines that you want to share with us before we leave?"

"No, you know everything I know, and even then some," replied Bill. "If I learn more about the murder of Mr. Koubiel, I'll let you know."

"That won't be necessary. You will be working with the detective on that piece of the puzzle," Dan told him. "We do want you to provide any information that will help us with our final report to the Board."

Mary could see that Bill seemed to be less attentive than in the beginning of the meeting.

After they left his office, Bill poured another two fingers of Tennessee whiskey. He glanced at his watch and dialed New York.

35

Newark, New Jersey

FRANKIE GALLO OPERATED two business offices in Newark where he split his time between his U.S. Lithium Corporate office, freely using his William Smith alias, and his former office where he conducted his technology business and allegedly his crime syndicate.

Mr. Gallo was a person of interest in the Koubiel murder. The police didn't have enough evidence to arrest or even hold him. Besides, his lawyers would easily convince a judge that U.S. Lithium Corporation was a valid company with valid intentions, especially when you look at the worldwide consumption of lithium to make lithium ion batteries. The opportunities to make money in this industry were tremendous for everyone.

The police could search and seize only the evidence related to Mr. Koubiel's activities relating into the murder. The search warrant was specific. They would love to find evidence of fraud, but they wouldn't be able to find it. So the teams were briefed on looking only for:

- Notes, address books, telephone directories, and calendars with conversations or meetings with Mr. Koubiel.

- Financial transactions like checks or ACH payments to Mr. Koubiel or to his vendors.

- Documents received from Mr. Koubiel including letters, contracts, leases, or any content regarding his work.

- Documentation to known underworld figures regarding times and dates, actions, surveillance notes, and content that could possibly point to a hired assassin.

Two surveillance teams were positioned at each office location. When Frankie was spotted entering the office, the teams immediately swarmed both offices and began their searches.

Jimmy LaFleur, Newark PD, led one team into his original place of business, and Paxton Schumer, NYPD, led the other team into Frankie's U.S. Lithium office. They had backup SWAT teams, just in case.

Conrad joined Paxton's team. These were not violent raids where targets would be arrested, handcuffed, and transported to jail. Both searches were done like any other search—knock, announce intent, serve search warrants, and begin looking for the specific items listed on the search warrants.

Conrad was directed to Frankie Gallo and said, "Here is our warrant to search these offices to gather information pertaining to the murder of Mr. Kevin Koubiel of New York City, New York. Do you have any questions?"

"May I call my attorney first?" Frankie replied as cool as a cucumber.

"You may call your attorney, but we will begin the search now. We also request that you come to our offices to answer a few questions. I presume that you will want your attorney in that interview."

The teams started their systematic search with desks, computers, filing cabinets, and all other areas like closets, waiting rooms, bathrooms, and any other places that could be accessed within the office space. At times Mr. Gallo or his attorneys would stop the searchers if they opened files pertaining to other businesses, reminding them of the strict limitations of the search warrants.

Back at the Newark police station, Frankie and his attorney were cooperative and pleasant during the interview. His attorney intercepted most of the questions so little "new" information was gathered. What information he did confirm was his majority share ownership in U.S. Lithium Mines Corp.—a matter of legal record. He knew Ruth O'Hara in the Brooklyn office, and he knew Kevin Koubiel from New York City. He claimed to not know about Kevin Koubiel's death. He did receive a call from Ruth about her not hearing from Mr. Koubiel.

"I told Ruth to continue processing the leases and documenting the BLM maps," said Frankie. "I told her that Kevin was a free-lance sales guy and that he had his marching orders. We run a low overhead operation that relies on people doing their jobs—no performance, no payola. Capici?" Gallo opened his arms in typical Italian gesture.

"But you were paying him about $12,500 per month," replied Conrad.

"Those payments were only the tip of what he could make," said Frankie. "I'm sure you will find his stock option plan in the files. If he does his job well, he stands to make millions. We operate like any other free capital enterprise based on incentives."

"Where were you on--" Conrad looked at his notes and continued, "Monday, the fifth of June at 11 a.m.?"

Frankie looked at his attorney, who nodded. "In the morning on that date, I was in this office, here working."

"What were you working on, and what calls did you make or receive?"

Again, he looked at his attorney who gave him the nod. "I don't recall all of the detail without my calendar. I generally take notes on the calendar. I don't remember calls I made either. I generally have very busy days, so it's hard to remember the details."

"Let me refresh your memory," said Conrad. "I have your office calls for the day." Conrad then listed all calls into and out of his office. "One call from a Las Vegas hotel came into your office at about 8:00 a.m. Does this help your memory?"

"Not really, because I wouldn't know where the call originated."

"We think you received a call from Mr. Koubiel because he was staying in that hotel at that time."

His attorney interrupted the question. "You don't have to answer this question." Looking at Frankie, and then up to Conrad, he added, "He already told you that he didn't remember calls on that day."

Conrad pursued the telephone calls for the entire day, into and out of Frankie's office. He wasn't able to get any real information directly from Frankie other than he spoke with Mr. Kevin Koubiel and Ruth O'Hara on a daily basis.

"Let's talk about another subject—people and shareholders of U.S. Lithium Mines Corp. Who are the shareholders of the business? And, what are their roles in the company?"

Again, Frankie and his attorney huddled before he replied, "Don't you already know who they are?"

"Yes we do," replied Conrad, "but can you tell me the roles of your colleagues in the company?"

"My title is executive vice president. My role is to secure lithium mineral leases in the four states. The president and CEO handles the capital side of the business and the VP of operations will handle the mining operations—when that begins."

Conrad continued his questioning for another hour or so but gained little new information.

36

New York Police Department

"ONE POLICE PLAZA," BARKED CONRAD to the taxi driver. Thirty-five minutes from his hotel to his seat in a small makeshift command post in the NYPD offices. Along the walls were computer monitors showing videos and audio recording screens at various locations. Paxton and a number of technicians were typing away at their computers and making adjustments in preparation for the events.

"She's already arrived," said technician one. "Should be no more than an hour or so for her to complete the interview."

Conrad replied to Paxton, "Do you have your men ready to go?"

"Yes."

"In both locations?"

"Hey, who do you think I am anyway," Paxton replied using his best Marlon Brando's Godfather impersonation.

Conrad relaxed and worked on the scone and coffee from his hotel. "Paxton, how do you like your coffee? Next time I'll bring you some hotel coffee. I think it's better that the stuff they have here."

"Black is just fine. I'd appreciate it."

While the interview was being taped, he looked at his notes and thought about the other possible suspects and motives.

Mr. Koubiel may have:

- Double-crossed U.S. Lithium by switching to Lithium America Corp.—if Frankie found out then he would be a prime suspect.

- ~~Worked undercover with the Feds.~~ No evidence.

- ~~Found out about the scam part of the deal.~~ No evidence of this. He was still trying to close the deal with Mr. Jackson in Rock Springs

- Was murdered by competition. Is it possible Frankie told some of his gangland friends about the deal? Would they then want to take over? Would they start their own company?

"She's just walked out of the building," said the same technician. "We have the entire interview on tape."

A few minutes later, one of the monitors lit up showing a number being dialed—area code 212. "He's dialing New York—direct line into Rajah Malani," said technician two.

"So much for following Mary's instructions," said Conrad.

Bill Anderson had just called Rajah to tell him about his visit from Mary and Dan.

"Rajah, the jigs up! I don't know how she found out, but Mary Johnson and Dan Duggan know about our little business venture, U.S. Lithium. And they know all the players in the company! They told me a Las Vegas detective will be in to see me soon. What do you know?"

"Listen Bill, you have nothing to worry about. Our little company venture is perfectly legal," replied Rajah. "Why would a Las Vegas detective be called in?"

"They said that Kevin Koubiel was murdered. Did you know about that?" Bill asked.

"We don't have anything to worry about because we're not involved with the mineral leases. That's Frankie's job. I'll call Frankie to see if they showed up at his office. Just keep quiet until we learn more," said Rajah. "You need to call Ruth to find out where she is with her work, and I'll find out from Frankie how to go about replacing Kevin."

Technician two chimed in. "Rajah is calling Frankie on his private number. He hung up after just five rings."

Technician one added, "Bill is calling the U.S. Lithium office in Brooklyn. No answer and no message. Now he's calling Ruth O'Hara's home number." Watching the monitor they could all see that Ruth had picked up and the conversation was being recorded.

The team listened to the new recording.

Conrad and Paxton instructed the technicians to continue their recordings and brief the FBI agent supporting the team. They both felt good that the FBI didn't try to take over the entire investigation, and they were grateful for the technical and personnel support they did provide. Getting eyes on the ground was a lot more difficult for an out-of-state cop working in other states.

37

Rock Springs, Wyoming

CONRAD SETTLED BACK IN HIS SEAT on the way to Rock Springs, Wyoming. He wasn't sure how to pick up the trail of the "black suburban" that followed Buck Jackson home north of Rock Springs. He had one idea about how he could possibly identify the players. It was a long shot.

He and Paxton hypothesized that Frankie decided to take over the U.S. Lithium Mines Corp. for himself. From the information they received from Buck Jackson they theorized that the scam would go down as follows:

- Secretly acquire exclusive lithium mineral leases from property owners in a four- state area.

- Promise big monies to each landowner.

- Renege on each deal after they file the leases with the BLM.

- Sell the leases to other lithium mining and pro- cessing companies at a big markup.

The business had all the look and feel of a legitimate mineral leasing company with low overhead costs. However, they couldn't think of a reason to kill Kevin Koubiel.

Conrad hypothesized that a Las Vegas mob group was trying to expand their operations into a market with huge growth potential. Only a small amount of funding would be needed to secretly secure the leases to build a portfolio in America that would be similar to a "cornering the market" tactic. Certainly the mob bosses of Las Vegas could divert a small amount of their cash enterprises to fund the venture—but who could they be?

Conrad met Buck for coffee at Applebee's restaurant in Rock Springs on Foothill Boulevard.

"Buck, I appreciate your help in my murder investigation," said Conrad. "I'm hoping to identify the people that followed you on several occasions—especially when you were with Mr. Koubiel. You look much better since my first interview with you in Las Vegas."

"I feel much better as the images of that day fade away. You mentioned that you would like to pick up the trail of the black suburban that followed me; is that correct?"

"Yes," replied Conrad. "I would like to walk through the events and when you saw them, to see if you recall any further details. I've also spoken with the Police Chief in Rock Springs. They have access to a psychologist to perform hypnosis to see if you may possibly recall any details in your memory of those events. Since you did see the vehicle from front and back, it may be possible to recall the license plates. If so, then we could pick up the trail again. Would you be willing to be hypnotized and questioned about those events?"

"Yes, I'm willing to help. How long will it take, and are there any ill effects from the session?"

"The psychologist told me the session will take about an hour, and there are no adverse effects from the hypnosis," said Conrad.

"She did say that hypnosis is like super concentrated rest or sleep. You will awaken refreshed. I tentatively have an appointment with her for later today, but I wanted to talk to you first."

Conrad spent the next half hour refreshing his notes by going over his original interview with Buck in Las Vegas. He outlined the events by date, location, approximate time-of-day, and the descriptions of what he saw. He would give these notes to the psychologist and review the specific details he was interested in reviving from Buck's memory.

"Meet me at Memorial Hospital at one o'clock in the main lobby."

Conrad escorted Buck to the special examination room assigned by the hospital—this room also had a couch and chairs. "Wow, this facility is new and modern." said Conrad.

"Yeah, Sweetwater County is well equipped and funded from all of the coal, gas, and minerals in the county," replied Buck. "The coal industry has had its ups and downs the past few decades, natural gas is new and plentiful, and trona is also mined here."

"Trona, what's that?"

"Well, it's a source of soda ash that is used in many products including glass, paper, detergents, and textiles," said Buck. "Do you remember the old ads on TV showing the Borax, twenty-mule team wagon? It's the same stuff and we have lots of it."

"And what about lithium?" asked Conrad. "Is it the new mineral, da-jour?"

"Yes, lithium is used to make rechargeable batteries, a hot commodity in the auto and aerospace fields. FMC Corporation discovered lithium in or near their trona mine a couple of years ago," replied Buck. "Now you know as much about it as I do!"

The psychologist entered the room and Conrad introduced himself and Buck to the young lady. "The police chief told me you are working on a murder investigation and would like to see if hypnosis will help clarify Mr. Jackson's memory. Is that correct?"

"Yes." "Yes." both Conrad and Buck Jackson replied almost simultaneously.

"Well then," she said. "Let me take care of the paperwork. Here is a consent form for your signature, Mr. Jackson. This is our standard practice. I will tell you that we have never had a single adverse effect from hypnosis, and we will not ask you any private questions. We will record the entire session for you to review afterward. Do you agree?"

"Yes I do," replied Buck.

"Then lets get started," she said directing Buck to a confortable chaise-type lounge, where she had him remove his boots to get comfortable, and drew the shades on the windows to darken the room.

"I am going to walk you through a series of questions that will help you relax and have your mind focus on my voice. You will always be in control of your body and you won't do anything you wouldn't do normally. Do you understand?"

"Yes," said Buck.

Buck reclined on a comfortable lounge chair as she walked him backwards in his memory of time. Buck nodded his head in response to her questions and after a voice cue she snapped her fingers. Buck's head nodded.

"You are now at the National High School Finals Rodeo in July. Do you remember?" asked the psychologist.

"Yes."

"Who did you see at this rodeo?"

"A man."

"What was his name?"

"Kevin. From the big city. Wore preppy shoes, jeans, and shirt."

"What did his face look like?"

"Clean shaven, blue eyes, and wide smile with dimples. Short sandy hair, receding at temples."

"What did he tell you?"

"Employer wanted to buy my mineral rights."

"Who is his employer?"

"Don't know. Wouldn't tell me."

"What else did he tell you?"

"I couldn't tell anyone about the deal."

"What did he tell you to do?"

"Gave me a cell phone to use. Keep secret. Would call on cell with instructions."

"Did he tell you his name?"

"Yes, Kevin Koubiel."

"Did you see him leave?"

"No we couldn't walk together."

"Did he ever call you?"

"Yes, two days later. Meet him at Penny's Diner in Green River."

"Did you meet him?"

"Yes."

"What do you remember about him?"

"Looks like yuppie. Name is Kevin Koubiel."

"What did he say?"

"Looked at a BLM map. My land. His company would pay $100,000 up front. Could be worth $2 million."

"What did you say?"

"Could not damage my water. Must drill away from springs. His company will be okay. Next meeting we finalize."

"Then what did you do?"

"I drove home."

"Did you see any vehicle following you?"

"Yes.

"What else?"

"A black suburban with tinted glass."

"Do you remember anything else about the black suburban?

"Slowed down, let it pass. Saw driver and passenger."

"Describe them."

"Passenger was a white man with short black hair. Back windows were tinted."

"Then what?"

"Slowed to let suburban in the right lane."

"Did you look at the license plate?"

"Yes. A Nevada plate."

"Could you make out the number?"

"I think so. AY-0296."

"What else do you see?"

"At my exit, black suburban in front. Saw bent rear bumper. Turned right at light.

"Did you follow it?"

"No. Went straight to my ranch."

"What else did you notice?"

"At my road, saw the same black suburban drive by."

"How do you know it was the same suburban?"

"Same bent rear bumper."

"Take a big breath of air and slowly let it out." She paused, "You are now in Las Vegas, staying at the Trump International hotel just off the main drag. Do you remember it?"

"Yes, a tall gold building."

"Why were you there?"

"To meet Kevin about mineral lease. Would have all details of deal."

"Where did you meet him?"

"At Johnny Rockets. In Harrah's Casino, in back."

"Then what did you do?"

"Sat at small table. Looked for him."

"Then what did you see?"

"Kevin walking between the slot machines toward me."

"What did you do?"

"Stood up to shake hand. Heard loud pops like firecrackers."

"What else did you see?"

"Kevin falling forward. Slow motion."

"Now back up to when you first saw Kevin. Look at the still picture frame. What do you see in picture?"

"Kevin between the slot machines. Smiling."

"Do you see anyone behind Kevin?"

"Yes. A man."

"What does he look like?"

"Dark hair, short. Can't see his face.

"Go forward to next frame. What do you see?"

"Kevin reaching with hand. I see a man's face over Kevin's shoulder."

"Describe the face."

"A round white face. Dark eyes."

"What do you see in next frame?"

"Man's arm raised. Holding gun."

"Have you seen him before?"

"Yes. Passenger in black suburban looked right at me."

"After you heard the shots and saw Kevin fall forward, what did you see?"

"Gun pointed at me."

"What did you do next?"

"Ducked gun."

"Anything else happen?"

"Something hard hit my head. Thought I would die, light faded away. Hear only noises."

The psychologist took her time to bring Buck Jackson back to the current time. "When I snap fingers, you will awaken fully rested."

Snap!

Buck took a few seconds to raise his head and realized where he was and what he'd been doing. "How did I do?" he asked.

"You were an excellent witness," said the psychologist.

"Yes you were, Buck," said Conrad. "You gave us some valuable information about the people who shot Kevin—including a license plate number from Nevada."

"Wow, I never thought it was possible."

38

Anderson & Smith HQ, New York

"Mary, you and I will complete our analysis of the data and prepare our initial conclusions to present to the management team. Once we get the team's buy-in, then we'll prepare our final recommendations. Are you ready to continue?"

"Yes. What do we do with we our new discovery about the U.S. Lithium Corporation, and how it connects with Rajah Malani and Bill Anderson?" she asked. "We also learned that this company is linked to a New Jersey mobster, thanks to Mr. French."

"Let's analyze this new information later to see how it may impact our conclusions," Dan said. "I suggest we divide and conquer the analysis of all our data."

"Sounds good."

They worked the remainder of the day reviewing all the information from the interviews. Then they started the arduous task of looking for common themes about what the information was telling them. Grouping into common themes, they were able to make sense of the information. Conclusions began to emerge for the first issue: Will the technology work?

They worked past the normal office hours when Dan finally said, "Mary, I don't know about you, but my mind is turning to mush. I'm not sure I can objectively analyze anything else tonight. How about you?"

"Me too. I'd love to get a fresh start in the morning."

They both straightened up their work papers. When she looked up at Dan cleaning the white board, she suddenly felt flush, "Dan, I really like working with you. Thank you." Looking him in the eyes, she said, "Would you be up for dinner with me tonight? I would like your company."

He looked into her beautiful green eyes, suddenly stricken with a wave of emotion. "Yes I'd like that."

THEY WALKED TO LA FONDA DEL SOL, in the Met Life building, taking less than five minutes. The food was excellent Spanish tapas—perfect for after work, and the décor great for private conversation.

"Are you holding up okay?" Dan asked. He knew she endured a lot of stress during the interviews, especially in St Louis, with the shock of sitting next to Conrad French on the plane home from the West Coast, learning about AeroStar's CEO Rajah Malani, and the exposure of underworld figures. "Let's talk about it."

They waited for the waiter to pour two glasses of Tempranillo. She thought before she spoke, "Dan, thanks for being a good friend and mentor." She paused a long moment and looked into Dan's brown eyes, "I really liked working for you."

Dan waited to see if she would continue before he replied. "You're welcome. Yes, you are a great work partner and confidant."

They sat quietly sipping their wine. Dan watched her face and expressions carefully as the wine let her face glow and her eyes soften. Her shoulders and body relaxed. She took another sip and looked up at his eyes again.

"I feel like I can share anything with you—what I think, how I feel," she said.

"How do you feel?" she asked softly. "You are a good looking guy. Do you say this to other female colleagues? Do they all become best friends?"

"Mary, please don't say it like that. I try to maintain a professional working relationship with you, and not cross the line. That is how I always work with you and other women—colleagues and clients. But if you think I am crossing the line, please tell me."

Her defense mechanisms suddenly dropped. "I'm sorry, Dan, for saying that. No, you haven't crossed the line. I'm having a difficult time working with you, and . . . find myself thinking of you as more than a friend. I think I'm falling for you . . . does this make any sense?"

"I think we should be good friends. But I'm not sure we should try to take it further. Not now, anyway, since we still have a lot of work to complete." He reached across the small table and took her hands cupped into his and squeezed them.

Her heart racing, she looked into Dan's eyes and squeezed back. "I'm glad you are my best friend." She smiled. "And mentor too. I feel much better now. This project has raised my emotional senses. But I do look forward to completing it with you."

THE NEXT MORNING, DAN WAS IN THE OFFICE EARLY. He didn't sleep well because he couldn't take his mind off Mary. *She's beautiful. She's smart. She's great to work with.* And, she offered ideas that challenged his thinking.

Mary arrived just as Dan finished writing his notes on the white board. "I hope you are ready to finish our analysis," he said. "Mary, about last evening . . ." He held up his hand to stop her from speaking. "I just want you to know that I really enjoyed being with you, and that I am feeling fond of you also. I hope you understand that I'm not rejecting you."

Mary replied slowly, "Thanks Dan. You know I really like you also. But I agree that we can continue working together without having personal relationship issues get in our way. I will contribute my best so we can get through this engagement. Thanks for listening to me last night. I really value your help and our working relationship."

"Now that we've done all the hard work, let's have fun with this last piece of the puzzle. Okay?"

"Yes . . . Okay." she replied.

Mary sat at the conference room table and spread out her work files, "I have all my interview notes and summaries. Just give me a few minutes to organize them here on the conference table."

"We have a lot of data to analyze, and I trust we'll get our interim report ready for the buy-in review with the management team."

Dan continued, "Let's start with a discussion of our general thinking about what we've learned—before we align the hypotheses. I'll start with my general observations, it you don't mind."

"No go ahead."

"We knew from the beginning that Rajah was trying to steer us and limit our investigation. He wanted only a favorable report for the Board. Then Bill Anderson at ComStar gave us a runaround story also. It wasn't until we gathered actual data on the DSR from Santa Susana that we leaned about the cobalt alloy issue."

"I agree. And with the DSR design, its success is dependent on the cobalt alloy magnet in the nuclear generator. The rare earth metals seem to be in short supply—at least for the necessary quality," said Mary.

"The engineers we met all believe the technology is available to build a hypersonic spacecraft," Dan added. "They said the only missing piece is a rocket with enough thrust. I see the only flaw in the DSR plan is the nuclear generator. If the "NucGen" doesn't work, then the whole plan won't work."

"The problem right now is the source and quality of the rare earth metals needed to make the cobalt alloy. And, the cobalt alloy may have a forging issue. It may be too brittle—this potential issue has not yet been resolved," Mary added.

"I believe the problem they must resolve quickly is the source and quality of the rare earth metals," said Dan.

"I agree," she said. "But what do we do about the corruption or independence issues with Rajah and Bill?"

"Yes, what to do about U.S. Lithium Mines Corp. and Rajah and Bill," Dan repeated. "I'd like your input from a human resources perspective? Then consider what a prudent Board of Directors would do if given this kind of conflict at the senior management level? Lastly, what laws did they break? We'd need some legal advice for that. Let's proceed with the issues we've prepared for. I'll have Robert take the legal question to someone in the firm with this experience."

They went back to the white board to continue with the conclusions and recommendations at hand. Dan asked, "What can we conclude from the findings we've identified?"

Mary thought for a moment. "I think that we can conclude that the entire project of a SSTO spacecraft is possible; that the critical technology for this spacecraft is the DSR; that the nuclear generator is still questionable because of the quality of the rare earth metals."

"Very good," Dan said. "I think this is exactly what we communicate to the Board. So what would be our recommendations to the Board?"

"We need to find out what the problem is with the rare earth metals. We need someone go to India or Afghanistan to find out exactly where they are getting the metals, the availability, and quality of the assays. We also need to know how much will be needed to build the prototype of the NucGen," said Mary

"I agree. You nailed it. Let's wrap up our report that we take to Rajah to get their buy-in."

Dan loaded up his PowerPoint presentation that he created at the beginning of the engagement. The Issue-Based Technique (IBT), when used during the proposal phase, is a perfect guide for the final report.

39

Port St. John, Florida

Bɪʟʟ Aɴᴅᴇʀsᴏɴ ᴀʀʀɪᴠᴇᴅ ɪɴ Pᴏʀᴛ Sᴛ. Jᴏʜɴ, Fʟᴏʀɪᴅᴀ, on a chartered jet to attend a special meeting called by Harold Zaben. Bill went directly into Harold's private conference room where Raymond Dabler was already seated.

"Can I get you something to drink?" asked Harold.

"No thank you. I had plenty of coffee on the flight down."

"Did you see this news release about the SpaceX recent failure?" asked Harold as he handed Bill a copy of the article he just printed from the internet. "NASA just lost a satellite valued at $112 million riding on a Falcon 9 rocket. SpaceX has already negotiated a settlement to NASA for future cargo launches to the International Space Station," he added.

"I didn't know that. Add the rocket, fuel, and launch costs to the payload cost; this was a costly failure for SpaceX," said Bill.

"Yes it was." said Harold. "I want the three of us to talk about our options going forward with the Aether Program. Specifically focusing on our business approach, since SpaceX's recent launch failure. I know it's premature given that we still need Board approval for the spending. Bill, I asked you down because I believe you know the latest status of the spacecraft design. We have a handle on the propulsion side of the project—the DSR. I want to know what you think we can do to move the project along, or what we can do to start a dialogue with NASA? What do you think?"

"I believe we are on target per our original project plan for the spacecraft," replied Bill. "How about the DSR prototype?" he asked.

Raymond Dabler jumped in, "We are perplexed with the Anderson & Smith consultants and are awaiting their report. They have been digging into the NucGen design and the Nelson ion thruster. They looked into Chloe's design and prototype of the Nelson ion thruster and Lizzy's nuclear generator prototype. Both Chloe and Lizzy claim to be on target per our plan. I did learn that the consultant, Mary Johnson, stayed an extra day to interview Zeke Kilbaggon in our production test facility. Zeke tells me that he's waiting for a sample of the cobalt alloy so that he can form it into the prototype core. He told Mary the same."

"Bill, what's your take on the consultants' work? You and Summer spent some time with them," said Harold.

Bill thought a moment before carefully answering, "They spent time talking to engineers who've worked on other designs like the X-15, the X-30, and the Boeing SST. They called this 'external research' to help support the feasibility aspect of the design."

Harold continued, "Thanks Bill. I think we should work on a strategy to get NASA looking at us again. We'll have to tell them about our new technology of a low-energy launch SSTO vehicle. I suggest we approach them for initial funding sooner rather than later. I think that they may be willing to fund us through proof of concept."

"What if our NucGen plant fails? Everything we are doing is dependent on the success of starting the Nelson ion thruster. Without the NucGen the whole project fails," replied Raymond.

"I don't see that happening," said Harold. "We all saw the cobalt alloy test and the specs to build the NucGen. I'm confident."

"And what about our rare earth metal supply chain?" asked Bill. "I don't see the consultants going to India or Afghanistan to validate the supply."

"Let's answer one question at a time," said Raymond. "I think we work on a pitch to NASA to secure funding now to complete the prototypes. I think that NASA needs and wants competition, rather that putting all their eggs into the SpaceX basket. That will take the pressure off our plan to spend our own money up front."

The three worked on the pieces of the plan that would be needed for the Board to approve. They all agreed to call the project "Hermes" after the Greek god of speed—speed was necessity for success.

BILL ANDERSON LEFT PORT ST. JOHN via the corporate jet and flew directly to New York's LaGuardia. "How was the flight?" asked Rajah, being polite.

"Before I get into our new plan to beat SpaceX, I want you to tell me what the outcome is from the murder investigation by the Las Vegas detective. How do you plan to keep the lithium company secret from the Board?" asked Bill.

"I called Frankie Gallo. He told me the FBI gathered files from his personal office and our business office in Brooklyn in their investigation of Mr. Koubiel's murder. Frankie was interrogated by a Las Vegas detective and the FBI. He did have his attorney present. He told me that he didn't know anything about the murder and that he wasn't involved," said Rajah.

"Did you make any plans with him? What are we going to do about Kevin's loss? How should we proceed?" Bill asked with a tone of urgency. "I don't know what to do."

"Relax, Bill. I think I have a 'go forward' plan for U.S. Lithium," said Rajah. "First we do nothing until we find out who murdered Kevin; then we plan our options. I think the options are: one, close the operation; two, hire a new salesman; or three, sell the business. We can rely on Frankie to help us decide the best way forward given his contacts. I would like to distance myself from him because of his underworld connections now that the ownership has been revealed."

"But what about our jobs with AeroStar? Don't you think the Board may fire us?" asked Bill.

"Here's what we do. If the Board calls me on the carpet, I will simply tell them that all is legal and above board. Stealth and secrecy was important for Annokkha Drat Exports to get a foothold in the lithium mining business. Annokkha Drat is a minority owner to keep them under the radar. Annokkha Drat is funding most of the mineral leasing phase, and you and I are the brains behind the business. Frankie has the operational skills to acquire the mineral leases, and he told me that this was his plan for exiting his underworld operations. I'll give the Board our business plan that shows the financial gains that Annokkha Drat Exports and AeroStar would gain from the business."

"What do you think they will do about not getting their approval first?"

"We really didn't need their approval because this is Annokkha Drat's plan and we are only facilitating the U.S. operations effort."

"Don't you think they will challenge us for conflict of interest?"

"I wouldn't call it conflict of interest; perhaps we've been unscrupulous or canny. I don't think we did anything illegal. Once they see the financial upside to this business venture, they will forget about the conflict of interest. Perhaps give us a slap on the wrist, if you know what I mean. Maybe they will give us a bonus for such a brilliant plan. All in all, I believe we can control them. I've started drafting a plan to be proactive on this, but I'll need your help. Bill, will you help?"

Bill thought about Rajah's ideas then said, "Yes, I'll help."

"Good. Now tell me about your meeting with Harold and Raymond," said Rajah.

40

AeroStar HQ, New York

Rajah Malani walked into his private conference room at exactly 2 p.m. His leadership team on the Aether Program—Parker Jones, Wade Williams, Bill Anderson, Harold Zaben, and Raymond Dabler—was seated at the conference table. "Hello to all of you and thanks for being here on short notice." To his knowledge, only Bill Anderson knew about the consultants' discovery of U.S. Lithium Mines.

Rajah continued, "I believe you all know about what's been happening at SpaceX after their recent launch failure. The reason we're here today is to finalize our plan to move our Aether Program timetable forward. Here's an outline of where we are and what we intend to accomplish." Rajah handed out the agenda for the meeting.

"The purpose of today's management meeting is to reach consensus on initiating a new project called the Hermes Project—Goddess of Speed. Hermes Project will speed up the timetable for negotiation with NASA about our alternative rocket delivery vehicle superior to the failed SpaceX offering," he continued. "Here are three actions for discussion: 1. Close out the Strategy Review Engagement; 2. Show NASA proof of concept for our new Spacecraft; and 3. Show NASA the economic benefit of our spacecraft delivery technology." Rajah waited a few minutes as the team reviewed his agenda.

"As you all are aware, our Board of Directors must have confidence in our strategic initiative before they will fund the Aether Program. We're in the final stage of the consultants' engagement. We still have some proof of concept research and development before we can deliver a prototype system. I would like your input about what actions could we, or should we take to successfully complete the consultants' report," said Rajah. "Let's go around the table and each of you provide your input. Starting with you, Harold."

Harold drew in a deep breath before he started, "First, we must know what the consultants' report says."

"Yes. Bill and I will get a preview of their recommendations next week," said Rajah.

"Excellent. Then your job will be to make sure the recommendations are convincing to the Board. I hope you and Bill can do that," said Harold. He nodded to Bill and Rajah.

"We think the consultants may have a concern with the NucGen development including the material sourcing. Am I correct about that, Raymond and Bill?" asked Harold. Bill and Raymond nodded.

"With that said, then we need to resolve the material issue quickly. I suggest that Raymond take this one and work with Lizzy at the Santa Susana labs to understand if there are any potential delays. Even if there may be a delay, get it quantified so that we can still convince the Board."

"Thank you," said Rajah. "Wade, will you facilitate the discussion on the second agenda item? I think it's fair to say that we need to paint a picture to NASA."

Wade Williams took over the remote control to continue his portion of the discussion topic. "We think its time to start discussions with NASA about our new advanced technology in our DSR—the Nuclear Generator and Nelson ion thruster. We can show them our new reusable spacecraft and launch profile. We have enough material to impress them about the possibilities of our technology."

Rajah facilitated the last agenda topic: show NASA the economic benefit of their spacecraft delivery technology. "We have a strong economic analysis showing how we will achieve less than $500 per payload pound. We believe this will be the deal closer."

Rajah then assigned each executive in the room actions to complete, and they agreed to meet in a virtual conference in one month to check progress. They also set up a team room on the company's website with VPN security, so they could capture and save documents showing the "value propositions" of AeroStar's new technologies that would benefit NASA.

After everyone left the conference room, Rajah approached Bill. "Will you please step into my office before you leave?"

When they were alone, Rajah asked, "What have your heard from anyone regarding U.S. Lithium?"

"Nothing. What about you?" replied Bill.

"I think we are good for now. If needed, I'll use our 'stealth and secrecy' excuse to show I'm on top of it all," replied Rajah. "I believe we will become the largest producer of gadolinium from our mine in Afghanistan, and with our plan to secure lithium mineral leases in the United States, we will become a world supplier in the future. Bill, I need you to continue your support as we get this plan rolling."

41

Upper East Side, Manhattan

Mary's commute from the office and bought three colas and packets of peanut butter crackers midtown to her studio apartment on the Upper East Side took less than thirty minutes door to door. Tonight she was tired after working with Dan on the presentations for AeroStar. *One more day and we should be in good shape for the buy-in discussion.*

"Hello, this is Mary."

"Mary, I'm so glad I caught you. This is Conrad French from Las Vegas."

"Hi, Conrad. It's good to hear your voice. How is the investigation going?"

"That's the reason for my call. Do you have a few minutes to go over the investigation with me?"

"Sure. Go ahead," she replied.

"I'd rather do this face to face," he said. "With all of your help, I'd love to buy you dinner."

"Are you in Manhattan now? I just got home from work and I've had a really long day," she explained. She wondered what the urgency was.

"Mary, this is my last day in New York and I'd really like to thank you personally, but I do understand," he said in a rather

189

disappointed voice. "Perhaps the next time I'm in New York we can get together."

Mary thought a minute then replied, "Conrad, I know a really interesting eating place called Eli's Night Shift. I'll meet you there in about twenty minutes. It's not a dinner place, but they do have good food and drinks."

"That sounds great. I'll be there in twenty minutes or less," exclaimed Conrad.

Twenty minutes later, Conrad arrived at Eli's and got a table for two. A few minutes later Mary arrived and gave Conrad a big warm smile and a friendly hug.

"Good to see you," said Mary.

"Same here. I'm really glad you could see me on such short notice."

"Conrad, you've been my mentor and confidant on a serious murder investigation, and you cannot leave New York without telling me the latest status," she said.

After ordering two drinks and looking over the menu, Conrad began his tale of the Kevin Koubiel murder. "I investigated Rajah and Bill at AeroStar, but they had solid alibis. I investigated Il Capo and after a thorough assessment came up empty handed. I even went to Rock Springs, Wyoming, and had a psychologist hypnotize Buck Jackson. We were able to get a license plate of the vehicle that followed him in Wyoming, and his subconscious memory was able to match the killer's face to the vehicle driver." He paused to take a sip of his beer. "I did some old fashioned detective work to track down the black suburban in Las Vegas to get the name of the driver.

"Buck Jackson wasn't able to recognize the face from my picture line up, but I eventually found the hotel clerk that recognized him on the days he followed Buck. I connected all the dots and made an arrest. I learned that another mobster who knew Il Capo decided to get into the 'lithium minerals' business to push Il Capo out," Conrad said.

They chatted during the meal, and then Mary told Conrad she had a long day of work tomorrow.

"I do want to thank you for helping me with my consulting engagement," said Mary. "We're in the process of preparing our final presentation to the Board of Directors. We will most likely tell our sponsor about the additional findings of the U.S. Lithium Mines. Rajah and Bill will have to explain their actions to the Board. While it doesn't change our recommendations, it may affect Rajah and Bill's future in the company."

"Mary, you are really a special person, and I just want to tell you that I've grown fond of being with you—whether it's work or pleasure. If there was any way we could remain friends, I'd welcome the relationship," replied Conrad in his smoothest and most confident voice ever.

"Conrad, you are special too. Yes, we can remain good friends, but I must confess that in my new job, I don't plan to get romantic with anyone for a while. Let's keep in contact. Please let me know how your case is going. Are you okay with this?"

"Mary, sure I'm okay," replied Conrad. "I believe we'll remain long term friends. I'll never forget the work we did together."

42

Kandahar Airport, Afghanistan

S ECURITY WAS HIGH WHEN POYA, the Fixer, arrived at the airport gate. The military guards asked their usual questions: "What is your name? What is your business at the airport?" Then they made him get out of his vehicle and open the trunk. One man walked around the car using mirrors to inspect the underside. Another guard inspected all contents in the trunk. They asked him to open the package that was wrapped by Abdul. Inside was a Styrofoam box containing three metal vials. The guard inspected each one and even shook them to hear the contents rattle—sounded like rocks. "What are inside these?" the guard asked.

Poya showed the guard the manifest as Abdul suggested. "These are mineral samples from a mine. I am courier and I do this delivery many, many times. Why you inspecting my car?" Good thing I left my pistol at home like Abdul requested.

"Not your business. You may pass."

Poya jumped into his car and quickly drove through. As he'd done many times, he drove to building 4400 where he delivered the package to the Federal Express agent.

"Please seal box because guards made me open," said Poya. The entire transaction at FedEx took less than five minutes, and Poya received a pink copy showing the tracking number and client codes.

As he pulled out of the airport gate, Poya looked at his instructions from Abdul. He'd given him another assignment to pick up a package in Kandahar and deliver it to Abdul at the mining facility. He pulled into the parking lot of a small building he'd visited many times, Afghanistan Imports and Exports. He knocked on the door to alert a male attendant.

"It's you, Poya," said the clerk who unlocked and opened the door.

"I'm picking up package for Abdul at mine in northern province," said Poya.

"Your name?" He always asked even though he knew him from previous visits.

"Mohammad Poya."

"Please come and sit here while I get package."

Poya looked around the small office while he waited for the clerk to return with a small sealed box.

"This for you. I understand you know where deliver to Abdul Wazir," replied the clerk checking a box on his manifest.

The Fixer was thinking of the jobs that Abdul had given him and was trying to figure what he was up to and why all the secrecy. They were all delivery assignments—pick up a package from Abdul, deliver to FedEx. These packages were always sent to a Mr. Bibi Gul Shinwari in New Delhi, India.

Poya thought about the transactions. Sometimes he delivered minerals to Abdul at the mine, and sometimes he picked up minerals from Abdul and delivered them to Federal Express, going to New Delhi, India. He wondered where the minerals he delivered to Abdul came from. Maybe he should learn more about Abdul's sources. Maybe that information could be worth money.

43

AeroStar HQ, New York

"Mr. Kavanah, Mr. Duggan, and Ms. Johnson from Anderson & Smith are here to see you," announced the executive assistant to Rajah Malani.

"Please show them to my conference room," replied Rajah. He was eager to hear their recommendations from the engagement, and he was also prepared to review a proposed change in strategy for the company—with or without the consultants' agreement. He needed to know the possible consequence of the U.S. Lithium Mines revelation. Rajah was an expert at making himself look good under any circumstance. He wasn't concerned with their knowledge of his side work in the lithium business.

Robert Kavanah set up his PC and connected it to the conference room system at the head of the conference table, he asked Mary and Dan to sit in chairs on each side of the table. He liked to have the client executive team sit interspersed with his team, creating the feeling of a team that worked in harmony. His goal was to get Rajah and his senior executives to buy in on the tentative recommendations.

A few minutes later, Rajah entered the conference room and closed the door. He sat at the opposite head of the long oval table facing Robert Kavanah, nodding to Mary on one side, then Dan on the other.

"Before we get started with the Aether Program Executive team, I have a question for you and your team," Rajah said to Robert. "I'm aware that your team discovered U.S. Lithium Mines Corporation during your engagement. Do you plan to include this finding in your report to the Board?"

Robert paused as he opened one of his folders and reviewed the top document. "Rajah, we plan to include this in our report, and we want to talk to you first. While this finding does not change our conclusions and recommendations, it may indicate a situation of potential conflict that may not reflect favorably on you and Mr. Anderson."

"I strongly oppose you reporting on this because it was not part of your engagement objective," replied Rajah. "Therefore I want you to remove any mention of the U. S. Lithium Mines in your report."

"Mr. Malani, we are an independent management consulting firm hired by the Board of Directors to answer the question, Will the spacecraft design work? We would be grossly remiss in our report should we not report this finding, and we would be subject to legal consequences from your Board. We want to discuss this topic with you before we meet with the other executives on the team." Robert pulled three handouts from his folder and handed them to Rajah, Mary, and Dan.

Rajah read the document and his face hardened as he contemplated the words. He read a simple outline of their findings related to the murder of Kevin Koubiel and the ownership records released from the Delaware court. He scanned down to the conclusion about U.S. Lithium Mines.

"Conclusion: Success or failure of U.S. Lithium Mines Corp. will not affect the success or failure of the Aether Program."

"Rajah, you may want to present your own justification for the creation of U.S. Lithium Mines Corp. to the Board," said Robert. "Your company's business practices for this engagement are not our concern at this time. Nor is any perceived management discretion unless it relates directly to the spacecraft design and whether or not it will work."

"I will think about this," said Rajah. "I have another urgent matter to discuss with you because it is extremely relevant to your final report." Rajah pulled out documents from his folder and handed them to Robert, Dan, and Mary.

"We plan to initiate a new project named Hermes—goddess of speed. We believe that your firm will help us succeed." Rajah paused for effect. "Due a recent Falcon 9 rocket failure of SpaceX, we want to take a proposal to NASA now so that they know we have a viable, alternative solution to SpaceX. We think NASA may act by funding our prototype program, rather than waiting until we demonstrate the prototypes."

After a few minutes of quiet while the consultants read Rajah's document, Robert replied, "I think this may be a valid strategy to take to the Board. Dan and Mary, what are your thoughts?"

Dan said, "This early move could put your long-term strategy in jeopardy. Rather than sole source the technology development, NASA may want to get proposals from other players in the marketplace."

"Good point, Dan," said Robert. "Rajah, do you believe your ten years of R&D in the Nelson ion thruster and the Nuclear Generator could be leveraged to secure a sole-source deal?"

"We are at least five to ten years ahead of any other company on this technology," replied Rajah, "so however we pitch this to NASA, we should make this point very clear."

"How do we satisfy the Board to move quickly when they themselves have doubts about the technology?" asked Mary.

"May we review your proposed recommendations to the Board?" asked Rajah. "Then maybe we can satisfy both efforts in one proposed action."

"Rajah, do you want to see our proposed conclusions and recommendations now, before we include the management team?" asked Bill Kavanah.

"I think that if we come up with a plan that satisfies the Board and gets us talking to NASA sooner, then let's work on this now before we show it to the management team," he replied.

With the nods of Robert Kavanah and the team, Rajah picked up the conference phone and dialed his assistant. "Will you please let the Aether Program Management Team know that we won't be ready for them for another hour."

RAJAH AND THE CONSULTANTS took a short break and refreshed their coffee and water glasses.

"Dan, please take Rajah through our conclusions and recommendations," said Robert Kavanah.

"I'll be glad to," Dan said. He plugged his computer into the presentation system and paged down to the Summary of Conclusions and Recommendations page.

"As you can see, we set out to answer the question, Will the spacecraft design work? Our conclusions to the issues are shown here."

Dan read the long list of issues he knew from the proposal and Rajah nodded. Then one by one he showed the findings for each issue.

"We've concluded that the source and supply of rare earth metals is unknown to date, and hence we do not know if the supply chain is adequate. This means that we cannot answer whether or not the Spacecraft will work because we do not know if the Nuclear Generator will work. We recommended that AeroStar conduct an independent review or audit of the supply chain for rare earth metals coming from Annokkha Drat Exports. We still do not have the assay report from your mine in Afghanistan, so we don't even know the type, quality, and quantity of the metals." Dan paused to let Rajah think.

RAJAH AND THE CONSULTANTS WORKED for the remainder of the hour to revise the final report. They added two additional conclusions: 1) AeroStar has a ten-year lead in nuclear generation and Nelson ion thruster technology. 2) The recent failure of the Falcon rocket used by NASA may be a valid reason for approving the AeroStar SSTO proposal.

Then the team modified recommendations to include that "AeroStar should initiate negotiations with NASA as soon as possible."

Action plans and next steps:

- Investigate the rare earth metal quality and supply by sending consultants to Annokkha Drat Exports to conduct a review of rare earth metals.
- Complete a proposal to NASA to secure funding for Aether Program.

AT THE END OF THE HOUR, Rajah called his assistant. "Please let the executive team into the conference room."

Rajah started the meeting by welcoming his management team and announced, "The good news is that the engagement results will support our plan to accelerate our pitch to NASA. The result we expect is to get funding to finish our prototype of the spacecraft. We could be replacing SpaceX within the next three to five years as NASA's prime contractor!" He went on to present the additional recommendations from the consultants.

44

Anderson & Smith Offices, New York

"Good morning, Rajah. How did your management team meeting go yesterday?" said Robert Kavanah. "Did the management team agree with the changes we made to the final report?"

"Robert, yes the team liked additions and changes you suggested, thank you. We all agreed to send your consultants to India as soon as possible so that we can close out that issue about the supply of the rare earth metals. How soon can Dan and Mary plan and coordinate that effort?"

"Dan and Mary are finalizing our report and presentation as we speak. I'll tell them to hold the report until they return with their findings so they may include the conclusions and recommendations in the final report version."

"Excellent. Have your team work through Bill Anderson to coordinate travel, lodging, and interview schedules at our Annokkha Drat Exports headquarters in New Delhi," replied Rajah.

"I will and get back to you, good-bye," replied Robert. A few minutes later, he dialed Dan and Mary in the conference room.

"Dan, can you put me on speaker?" he paused. "We have a change of plans. Rajah just gave us approval; you and Mary are going to New Delhi, India, as soon as you can get your visas," said Robert. "Please coordinate this trip with Bill Anderson to arrange for hotel and schedule of interviews. Get your passports to our travel office so they can expedite the necessary visas for entry to India. Meanwhile, you and Mary prepare your list of hypos and key questions for your interviews."

"Do you think we should travel to one of the mines to see the operation?" Dan asked.

"I'm not sure. Why don't you review your data needs and key questions with Bill Anderson, and then ask him if you need to see an operation. You know, we never received the assay report from India, so put that on your list."

"Robert, why do you want me to go too?" asked Mary.

"I want both of you to take notes and then compare them so we don't have any misunderstandings on what was said. You may even want to record your interviews, but run that by Bill first. We don't want to offend their culture."

"So why the urgency?" asked Dan.

"They want to expedite the final report so that the Board will approve the plan to proceed with the new plan for pitching NASA."

THE NEXT DAY DAN AND MARY completed their list of key questions.

"Mary, I think we're in good shape. I'll send these questions to Bill Anderson and we'll get him on a conference call tomorrow morning," Dan said. "What thoughts do you have?"

"Will we need a translator or someone to help us ask the questions?"

"My experiences working in India have all been great. Most of the people speak English well, and if they don't then we'll have someone sit in on the interviews to help," said Dan. "But let's include this question when we talk to Bill tomorrow."

"I'm quite nervous about going to India and worried about liking the food. I hope I won't offend anyone."

"You'll be just fine, so please don't worry. I've seen you eat Indian cuisine before here in New York. It's even better over there. And, if you really don't like it, the hotel will make something you like."

"Should we call Lizzy in Santa Susana to ask her about gadolinium and what we should look for during our visit?" asked Mary. "Do you think there would be anything that we can see to give us a clue about why her samples have varied so much?" she added.

"Let's give her a call and find out," said Dan.

It was mid morning in California when the telephone on Lizzy's desk rang. "Elizabeth McKibbin, may I help you?" she answered.

"Hello Dr. McKibbin. This is Mary Johnson from Anderson & Smith. Do you have a few minutes to answer some questions that Dan and I have?"

"Hi, Mary. Yes, I'd be glad to answer any questions you may have. Go right ahead."

Mary pushed the speaker button. "Hi, you're on speaker phone. Dan and I are working on our final report and we have some new developments. I'll let Dan explain."

"Hi, Dr. McKibbin. Mary speaks highly of you since she was in Santa Susana. We understand your situation with the gadolinium samples you've been receiving from India. We've received the go-ahead to visit Annokkha Drat Exports to see if we can find out more about the sources."

"I see. How could I help you?"

"Can you give us any insight as to what we should be looking for? What is the mining and refining process? What do the ore and the metal look like? Do you have any tips that would help us?"

"So you are trying to find out why I'm puzzled about the quality differences?" she asked.

"Yes, and we also need to answer the question about the size of the reserves—is there enough supply chain to meet your production needs?"

Lizzy thought about the request for a minute and said, "You are not geologists, so I doubt you would be able to identify the form of the raw minerals. Gadolinium is not found in its pure state because it is highly reactive to oxidation. It's found in several mineral forms, and you might be able to identify each one by color and texture. Will that help?"

"Would you send us pictures and documentation about the different forms? And, perhaps provide questions we can ask to help us discover the differences?" asked Dan.

"I think you will meet with Bibi Gul Shinwari in Metallurgy Research. He's quite knowledgeable, so have him show you the raw material samples that he used to create the finished samples he sent me. I'll send you the list of all finished samples I received and the dates," she said. "You can take pictures of each ore sample, then perhaps together we can determine the form of the raw ore. First, ask him to tell you the type of ore each sample was derived from. There are several types of ores. The two most common types are bastnaesite burundi, and gadolinite. But the best ore for mining is lepersonnite. It's very rare."

"This is great information. Is there anything you can tell me about the refining process? Is it possible that Bibi's refinement could cause the variations?" Dan asked.

"I think you should ask him this question and perhaps have him show you—if possible," she replied. "I think the differences are due to the ore source. Somehow, an impurity is getting into the metal sample I received. It had an unfavorable effect on my research.

"Bibi is getting his ore from their mine north of Kandahar in Afghanistan. He deals with Abdul Wazir who is in charge of requisitions and supply for the mine."

"Lizzy, thank you for working with us; we look forward to any information you may send us."

"Wow, look at this," said Mary. "Lizzy sent me pictures of the three mineral forms that contain gadolinium. They are beautiful crystals. I think if we can see the raw minerals that Bibi receives from the mine, then we might be able to understand any possible differences."

Mary then opened each picture and showed Dan. "I believe I can even recognize the mineral source based on these pictures. What do you think?"

"We'll just have to see what Bibi has for the samples he sent to Lizzy."

Mary printed the paragraph and the three pictures she received from Lizzy. She read the paragraph to Dan:

"Gadolinium is a constituent in many minerals, such as monazite and bastnaesite, which are oxides. The metal is too reactive to exist naturally. The mineral gadolinite actually contains only traces of this element. The only known mineral with essential gadolinium is lepersonnite. It is very rare."

Mary put the pictures in the research binder.

45

New Delhi, India

After more that eighteen hours of flight time from New York City to New Delhi, India, Dan and Mary arrived at their hotel at 10 p.m. local time. They crossed the international dateline so the day was still Tuesday in New Delhi. Going home, they would lose the day they just gained..

"Mary, it's best you go straight to bed. Your body clock will be a little confused so get as much sleep as you can. I'll meet you in the hotel restaurant at 8 a.m. tomorrow," said Dan.

"See you in the morning. I'm bushed," she replied.

Dan saw Mary sitting at a small table in the hotel restaurant and walked to greet her. "Mary, you look spiffy for a busy day at the office." Dan grinned showing her he was just kidding. But seriously, Dan couldn't help see how her slacks fit just tight enough to see her exquisitely shaped legs and thighs. She completed her outfit with black pumps, beige top, and wrap-around shawl.

"You told me to dress down so we won't offend our Indian clients," she replied. "I like your Dockers and plaid shirt. You look more like a Maytag repairman than a management consultant—just missing the nametag over the left side of your chest," she kidded back at him.

"I hope you slept well last night. Today we'll take a look at those samples that Bibi has and learn more about his sample refinement processes. I think it will be good to see their office and meet the people. I've asked the managing director, Gholam Sharma, to also allow us to have a briefing with their staff before we leave. This will go a long way to strengthen their working relationship with AeroStar. Do you have any questions?" Dan asked.

"I have no questions. I'll look for any signs of deception from Bibi or anyone else. I'll give you the sign if I see anything worth exploring. What do you think we'll learn?"

"I'm not sure what we'll find. But my questions are intended to find out what could be the differences in the samples. I want to see the assay report for the mine and understand more about the minerals. Remember that India and other parts of the world are much less regulated than we are in the U.S. Assay reports are not required to stake a mine or even open one. After we get background information from Bibi, then we'll ask him to show us his raw samples for each one he sent to Lizzy. Let's go do this," said Dan.

FROM THE FOUR SEASONS HOTEL, the taxi ride took forty-five minutes to Annokkha Drat Exports headquarters. The traffic was heavy initially and thinned out as they reached a more industrial section of the city. At one traffic corner, cars had slowed down to pass two white brahma cows casually crossing the street. Down another street Mary saw one house with a large open stall next to the home, like one for a big RV. As the taxi slowly passed, Mary gawked at the large elephant eating hay in the stall. "Hey, Dan, look at that," as she pointed.

"They treat their work vehicles, in this case a working elephant, quite well," he said. "The government doesn't officially sanction a caste system. This family is an example of a working caste that is higher than the "untouchable," or lowest caste. They hire out their elephants to move heavy objects. The owner directs and controls the elephant."

"See those blue tarps over there in that field?" Dan pointed. "That is an encampment of the poorest of the poor, formerly called the "untouchables." Many don't have shelter over their heads. In the monsoon season, they have nowhere to go to escape the rains except under their tarps. Often you will see their children, sent out to beg for money or sell hand-carved objects."

"I can't believe the amount of poverty everywhere," said Mary. They continued their ride in quiet.

Once in the managing director's office, tea was served—a tradition held over from the British. They shared small talk with Gholam Sharma and Bibi Shinwari about their travels from New York. Because AeroStar holds only minority interest in Annokkha Drat Exports, Dan cannot disclose the strategic initiative project. He explained their objective was to explore the rare earth minerals, specifically the element gadolinium, mined by Annokkha Drat as part of a research project for the Vega Group.

"You and Ms. Johnson have full access to staff as you need," said Gholam Sharma in his best Indian English, with a heavy accent. "Our mission to be successful now and in future with AeroStar. I asked Bibi to support you 100%."

"Mary and I thank you for your support and hospitality," Dan said. "Before we get started with Mr. Shinwari, Mr. Rajah Malani asked that we get a copy of the assay report he requested recently. This is for the mine in Afghanistan that supplies gadolinium."

"We just receive report from company that did assay study," replied Gholam. "I make copy for you to take with you."

"Excellent. Thank you," said Dan.

A few minutes later they met Mr. Shinwari and exchanged salutations and small talk—and were soon on a first-name basis. "We are interested in your process of refining the samples. We understand that Dr. Lizzy McKibbin has received a number of samples, and there have been some differences. We would like to see your process to see if we can identify possible changes. Do you have any questions?"

Bibi replied, "No questions, thank you very much. I'm ready to show you everything you requested in your email. Shall we get started?"

They followed Mr. Shinwari to his office. Dan took the lead in their interview to learn how he communicated with Abdul Wazir. He then discussed how Mr. Shinwari handled the samples he received including the extraction of the gadolinium to create the delicate silvery metal. Mary observed and took notes.

"Mary, do you have any questions?" Dan asked.

"Dr. McKibbin gave me a list of the sample numbers she received from you. I would like to see each of the raw samples to document the type of mineral you received," she replied.

"Let me show you my laboratory," said Bibi. "I have some of the latest equipment so that we can analyze and evaluate samples from our mines."

Bibi was right. The lab was huge and spotless. Several lab technicians in white lab coats were busy attending to equipment in one end of the laboratory.

He directed Dan and Mary to a large wall cabinet that had dozens of pullout drawers. "Here is where I keep all samples from mines. Here in this section are samples from Afghanistan mine."

Mary showed Bibi the list of samples Lizzy received by date. "Let's start with the first three samples," said Mary.

He looked at the list and retrieved the sample drawers. Inside each drawer was a sealed container with the raw mineral and a sealed glass jar with the silvery metal held in a liquid of some sort—to prevent oxidation.

Mary took out pages of bright white paper with labels at the top—sample number and date—in a large font.

On each paper she poured out the raw minerals and next to it the jar containing the gadolinium. She used her iPhone to photograph each sample. She reviewed each photo to ensure quality, color, and lighting. Lizzy told her how to record each sample and asked her to email the photos as soon as possible.

When she was satisfied that the photos were clear, she took each set of granules and looked at them with a magnifying glass she brought. All the samples were finely ground and looked like brown dirt with little or no color. The magnifying glass allowed her to see small specks of color. She wrote down discernible colors on each paper sheet.

The first two samples had bright specs of yellow and green, the next three samples had some pinkish gray, and the last three samples had no discernible color.

Bibi walked them through his laboratory process for handling and refining each sample. Bibi said, "I receive about 1500 to 2000 grams of the raw mineral to produce 10 grams of the pure gadolinium. I show you the process and steps."

Dan and Mary took notes on all steps of the process. In all cases, the lab technicians performed the handling in isolation chambers where the minerals were handled by machine, remote access, or with gloved hands to prevent possible contamination.

Bibi wrapped up the session. Dan said, "I reviewed the assay report that Mr. Gholam Sharma gave us per AeroStar's request. I think it says that the minerals you are mining in Afghanistan are from the bastnaesite mineral. Is that your understanding?"

"Yes, that is correct," replied Bibi. "We are pleased that we have bastnaesite; it has better yield than gadolinite. We explore for the lepersonnite in our mines, but cannot find this grade of mineral in Afghanistan."

Dan thanked Bibi for his support and asked, "Do you mind if we write our notes using your conference room? We would also like to send emails to our other project team members, so would you allow us to also use your Wi-Fi? We would just need to know your VPN ID and password."

"I think we can let you. I will ask our computer manager and have him show you how to access the internet. Do you have other requests?" asked Bibi.

"No, that is helpful. Thank you."

Mary had prepared her notes to send to Lizzy in Santa Susana. Using her world clock on the iPhone, Mary saw that Lizzy was 12 hours and thirty minutes earlier. What she sent now would be in her email box for about six more hours before she would be in the office to read her sample report and review the pictures. Mary would check her email at the hotel tonight after dinner. She was curious to know what she thought about the raw samples.

Mary said, "Lizzy won't be in the office for another six hours to read my sample report. I'm not a geologist, but to my eyes the first two samples were truly different. What if they are samples of lepersonnite?"

"Both the assay report and Bibi confirmed that the mine in Afghanistan has only the bastnaesite mineral. How could the samples be different?" asked Dan.

"I'm not sure, but let's head to the hotel. Perhaps I should call Lizzy and wake her up," said Mary. "Let's not discuss this here, but back at the hotel, okay?"

In the company car on the way to the hotel, Dan's mind was working overtime thinking of all of the possible reasons why the samples were different.

"What do you think happened?" Mary asked.

Dan held up a halting hand and whispered in Mary's ear, "Let's not talk about this in the presence of our driver."

"Boy, I'm famished," she said. "What do you think of the food at the hotel?"

"The Four Season's food and rooms are great," said Dan.

"I like the buffet. That way I can sample new Indian cuisines I've never had before."

On the ride back, the driver took the same road as the taxi driver. Mary got to see the house with the elephant stall again. This time the elephant was gone and the floor of the stall was cleanly swept up with only one bale of hay in the corner and half of a 55-gallon barrel for water.

Back at the hotel, Dan said, "Let's meet in the business center and plan our call to Lizzy and brainstorm ideas about how the samples can be so different. Then we'll make the call."

Thirty minutes later, they sat at a table and connected to the hotel's ethernet to look at the pictures, documents, and to read emails. "Look at this," Mary exclaimed. "Lizzy has replied to our email. Doesn't she ever sleep?"

Dear Dan and Mary,

It is unusual that the first two samples have specks of color. The assay report said the ore in the Afghanistan mine is the bastnaesite mineral. This corresponds to the pictures and descriptions of the other samples.

I believe that you should visit the Afghanistan mine and meet with Abdul Wazir and do the same analysis you performed with Bibi. THIS IS TOO IMPORTANT TO IGNORE. We must identify the source of the first two samples.

46

Travel to Afghanistan

Dan and Mary had eaten dinner at the Four Seasons in New Delhi. They were both a little nervous about their change of plans to visit the mine in Afghanistan in the next day or so. Dan knew the company had taken the necessary steps to ensure their safety, but his concerns still gnawed at him about going to the Kandahar Mine. It's a four-hour drive to the mine and four hours back. Too much could go wrong.

They moved to the bar and were sipping on after dinner drinks when Dan said, "We should get our visas and travel docs tomorrow or the next day from the U.S. Embassy here. We should take ore samples with us back to New York. What do you think, Mary?"

"Yes, we should get samples," she said. "That way we will get an independent assessment on the minerals. I'm glad the Embassy is providing a U.S. military bodyguard and a satellite phone for the drive to the mine. Do you think we'll be in any danger traveling the back roads? Four hours of driving is a long way."

"When we get to Kandahar, we'll be staying in military offi-cer quarters for food and lodging. Our bodyguard will have a spe-cial satellite phone with all the necessary numbers to the embassy and a special forces commander in the area should we run into some problems," Dan told her. "Lets ask them when and how often we should call in. Our driver has been hired by Annokkha Drat Exports. They know who to hire."

Mary thought for a minute as she sipped her drink, "Let's call the embassy staff in New Delhi before we leave to see if they know the driver that was hired just to make sure he is reputable."

"I'll make a reminder so we don't forget to check these details with the Embassy tomorrow," Dan said. "If all the paperwork get's done, I think we will be in Kandahar in a couple of days. Are there any other details you can think of?"

"I'll use the maps we received to mark possible routes we might take from our lodging to the mine and back," said Mary. "And I'll mark important spots inside the base and in the town in case we get lost."

"You're being paranoid," Dan smiled to put her at ease. "After I call the embassy tomorrow, we'll go over our questions for Abdul Wazir using Lizzy's suggestions. Then let's take a break and do some sightseeing and shopping in New Delhi. We need to buy khaki pants and shirts, comfortable boots, and any sundries we need. The hotel will hold our suitcases here for when we return." He paused and added, "We have to do this right. I suspect we'll need to change our recommendations. AeroStar may enter a risky venture that could cost them billions in investment. Failure to produce a working spacecraft is not an option."

THEY RECEIVED THEIR TRAVELING PAPERS including an Afghanistan visa, U.S. Embassy letter of approval, airline tickets, and the purpose of their business meetings. They'd fly from New Delhi to Kabul, change planes, and fly direct to Kandahar. They'd arrive at 3:30 p.m. and go directly to their hotel, a former U.S. military bachelor officer quarters. It's wouldn't be fancy, but it would be well protected within the airport security compound.

"Mary, you look stunning in your khakis," Dan said on the morning of their flight.

"Cut it out, Dan. These outfits were your idea, remember."

"Don't you think we'll blend in with the locals?" he said, kidding her.

"Fat chance," she said laughing.

They boarded an Airbus 320 for the flight to Kabul. The interior was rather plain—no entertainment console for every passenger. They immediately opened the air nozzles and pointed them directly at their faces. Dan said, "Here we go into the wild, blue yonder to Afghanistan"

Mary smiled weakly and nodded. Her face was a little pale.

"Don't worry. Our return flight is already booked on a nonstop back to New Delhi. We've got five days to complete our business."

Shortly after takeoff, Dan relaxed and his mind went to his dad. "I never told you about my father," said Dan. "He was a Green Beret with experience on special missions in Afghanistan. I loved him. God, the war stories he told us. He always encouraged me to be the best I can be. When I was fifteen, I told him that I wanted to be a military pilot and fly combat jets."

"We are on a military mission right now," she whispered in his ear.

"I told him I wanted to learn to fly now," Dan said. "I had already looked into the requirements for ground school and flying lessons. I even saved most of my after school job money.

"He told me that it would be costly, and he said he would match me dollar for dollar. But, I couldn't let it get in the way of my college education. With his support, I passed my ground school in three months, and my flight training by the end of the year. He was so proud of me when I showed him my single-engine pilot's license. I was really proud too."

"It's great how your mom and dad encouraged and supported you while you were so young," said Mary. "I thought you had to be a certain age to get a pilot's license."

"Yeah, they did. When I was seven, they enrolled my brother and me in karate. Mom took me to lessons for years—I even received my black belt. My brother and I were good athletes. We lettered in baseball and football. Looking back, I really believe our achievements in our younger ages gave us self-confidence and high self-esteem."

Mary looked at me, "Now I know why you are such a great guy. I feel safe traveling to Afghanistan with you. When I grew up in Modesto, my dream was to marry and have two children."

"Why only two?" asked Dan.

"I have two sisters and three brothers. Dinner was like a major production and all of us kids were assigned special duties," she said. "Never again."

"That's tough. I never thought about it from your perspective. I ate dinner at my best friend's house a lot. He was one of six kids too. I couldn't believe the dinner table talk and dynamics. The father was amazing. He paid attention to every child at the table no matter how small. He'd let each one ask a question, one at a time, and he answered them," said Dan.

"My mom was my guiding light. Reading so many books opened my eyes to the world, and gave me the desire to learn and to explore. I chose college over marriage. Dan, I'm glad I did too. This job is challenging, exciting, and full of surprises. You've taught me to trust my knowledge and instincts."

She looked at me sitting next to her. "What are you smiling about?" she asked just as the airplane started its descent into Kabul.

Dan looked at her as they hit an air pocket. "I thought of my father again. Our secret mission is like many a mission he encountered during his Green Beret career. Except we're not getting shot at. I wish he was here to see me now."

"No, our weapon of choice is our skill in getting the right information so that we can help our client. No shooting please."

"Remember, we're going to a hostile country, especially toward Americans. As long as we stick together, we will be just fine. We're dealing with company professionals, not enemy combatants," Dan said. "I know you will be able to catch any lies or deceptions, even though we'll be using an interpreter."

"I'll do my best."

The descent into Kabul was rather bumpy from the hot air lifting off the desert floor. The pilot made what Dan would call a military-style landing. The touchdown was firm without a big bounce, like a jet fighter landing on an aircraft carrier. Then the reverse thruster and brakes were hit hard, causing the nose of the aircraft to dip as the plane decelerated. Dan could feel his seatbelt holding him firmly in his seat. The pilot turned out at the second taxi way and slowly eased into the front of the terminal.

When they reached the exit door, a blast of hot air hit their faces as they walked down the gangway. A bus took them to the main terminal at Hamid Karzai International Airport. It was more modern than Dan expected—spread out wide with the air traffic control tower in the middle made of brick and glass. It reminded him of many U.S. terminals build in the 1960s, but this one was built by the Soviet Union.

Once inside the terminal, they queued up to show their passports and travel papers. Only three passport control agents worked the glass booths. They handed the agent their passports and embassy letters. The agent looked at the passport, and then back at them comparing pictures. He swiped the passport into the computer, looked at all the secret information they had on them, and asked, "What is your purpose for visiting Afghanistan and how long is your visit?"

They knew the right answers, "Business. We will be here five days."

Once satisfied, the passport control agent said, "Welcome to Afghanistan." He stamped the visa page showing the entry date.

Inside the terminal, they looked at the departure board to see their second leg into Kandahar and walked to the terminal. "That was pretty easy," said Mary. "I thought for sure we'd be given critical scrutiny."

Dan had already looked at the two Afghanistan airports on his computer to get as much information as possible. He liked being prepared—area maps, local languages, and culture. He even read up on the two official languages of Afghanistan, Pashto and Dari.

He wrote down a few words for *hello, goodbye, thank you, yes and no.*

AN HOUR LATER, DAN AND MARY BOARDED a Fokker F-100 regional jet. The second leg to Kandahar was a short flight, but was much bumpier than any flight they'd ever taken. The late afternoon heat rose off the desert floor causing major uplift and choppiness.

On the ground in Kandahar, they asked the information desk about transportation to their lodging in the bachelor officer quarters. A military Humvee drove them the few blocks to the BOQ building.

When they got out of the Humvee, Dan saw the look on Mary's face and said, "Okay, it's not fancy, but we'll be well protected within the airport security compound."

After they checked in, Dan asked the clerk, "Where can we get food and drink?" The clerk pointed down the hallway.

"Let's put our luggage away, then I'll meet you at the end of the hall for a drink," said Dan.

Mary nodded.

They met in the BOQ lounge, Dan said. "Not very fancy. How do you feel about on the ground inside America's longest war?"

"I feel really creepy," said Mary. "Like someone is watching me. I just can't put my finger on it." Mary sat across from Dan and held her arms across her chest, hugged her body, and looked around to see if anyone was watching her.

Dan ordered two glasses of white wine from the bar. Alcohol was allowed even though ninety percent of Afghani's are Muslim because Afghanistan is an American stronghold. Dan peeked at the bottle the bartender poured from—Gallo Chardonnay. *Not bad for being halfway around the world he thought. At least it's from California.*

"We are a long way from New York," said Dan. "We have a long day tomorrow and we should be well rested. Here's a toast to you and me. May we find the answer our client is looking for." Dan held his glass up to hers for a light clink of the glass. The wine was cold, fruity, and crisp on the pallet with a hint of oak. "Ah, this really hits the spot."

"Dan, thank you for all your support," said Mary. "You make all this hard work in strange places well worth the effort."

"Thanks Mary." The wine began to work its way through his body and mind giving him a euphoric buzz. His eyes welled up slightly as he looked into Mary's eyes, "You are wonderful. I can't tell you how lucky I am having you by my side."

47

Kandahar Mine, Afghanistan

AFTER BREAKFAST THEY PACKED TWO BAGS EACH and met in the lobby to meet the bodyguard. They each carried two backpacks with laptop and job folders containing the pictures of the ore samples from Bibi and the key questions to ask during the interview. Dan also packed a third backpack with water bottles, sandwiches, and snacks for their trip to the mine and back. He also brought his GPS device programmed with the coordinates of the mine, the Kandahar BOQ, and military security.

A few minutes before 6 a.m., a tall, burly Marine entered carrying an automatic rifle and a knapsack. "This must be our bodyguard," Dan said. He stood up and approached the Marine, who snapped to attention, "Master Sargent Randal Wilson at your service, sir."

"I'm pleased to meet you also," said Dan as they shook hands. "I want you to meet my associate, Mary Johnson from New York."

"Pleasure to meet you, ma'am."

"The pleasure is also mine," she said with a big grin on her face. "We are glad you are accompanying us on this journey."

The three of them gathered around a coffee table in the lobby as Randal spread out a field map that had yellow highlighted marks on the roads they would be taking. Randal gave them instructions on how he would handle encounters with local Afghanis.

"We should not encounter Taliban soldiers since they are in the north and we will be traveling south. But in the event we do, I will give you instructions on where to shield yourselves from possible fire. I am armed and have a satellite phone to communicate with the command post. I can have a helicopter dispatched to our location in a matter of minutes depending how far away we are," said Sergeant Wilson. "Do you have any questions?"

They looked at each other and shook their heads.

"Now where is the transport and driver?" asked Sergeant Wilson. "Unfortunately, your client arranged for your driver and vehicle. I would have preferred a Humvee with one of our drivers." Randal looked out the front door and saw a Ford Explorer park. An Afghani got out. "I think he's here."

Randal approached the driver, "Sob bakhir. Ma dorost dari yad nadaraom (I can't speak dari well)," he said, heavy on the English accent.

"You speak fine," replied the Afghani. "I am driver for Mr. Dan and Ms. Mary. I am company driver and go to mine many times."

Dan and Mary were right behind Randal and each greeted the driver. Dan said slowly, "Sob bakhir (good morning)."

"Good morning also," said the driver. "Is everyone ready? Let's get going. Long ride ahead."

They loaded into the Explorer—Randal took the front passenger seat, Mary behind him, and Dan behind the driver. "It is our policy to carry an automatic weapon," said Randal to the driver. "Where we are going, I don't expect any trouble from the locals or from Taliban."

"No problem," said the driver. "I do this many times."

They drove through Kandahar and everyone was quiet. Dan fiddled with his new GPS device, and Mary looked intently at the buildings and houses. Most of the houses looked even poorer than what she'd seen in India. There was only one striking difference: no people or vehicles could be seen at 6 in the morning.

Within minutes they were on the main road heading southeast. It's a two-lane dirt road and as far as Dan could see ahead, no other vehicles were on the road. The sun was already peeking over the mountains to the east casting a pretty picture of light beams spreading out from the circle of sunlight. When the light beams struck boulders and small ridges of rock, the low angle of the early sun cast long shadows. It reminded Dan of the black-and-white TV show *Twilight Zone*. The rapid staccato verse of music gave him goose bumps up and down his back as Dan wiggled in his seat—da da da da daaaaah.

The driver was driving fast on the dirt road. Good thing he's driving an American-made vehicle. Dan got his GPS working. On the screen was a little dot moving with no roads or details other than the surrounding terrain. The mine destination was a second dot on the map. Dan guessed Garmin hadn't mapped the roads in Afghanistan. Oh well, so much for technology.

As the sun rose in the sky, the temperature quickly jumped into the 90s. They kept the windows rolled up to keep the dust out, but the heat soared in the Explorer. The air conditioning didn't work.

Eventually the driver turned off of the main road onto a less used road that headed into the hills to the east. He slowed the Explorer considerably for better control and to keep the uncomfortable bouncing down. He knew the road like the back of his hand and expertly made the turns and negotiated rocks and the ridges of tire tracks.

After an hour of bumpy ride, Dan asked the driver to pull over so he could use the "facilities." Kandahar was no longer in view even though their elevation was about two thousand feet higher. They opened the doors to capture as much breeze as possible in a futile attempt to cool down the interior. They sat for a while drinking water and eating chips and pretzels. "By the way," Dan asked, "what is your name?"

"My name is Mohammad Poya."

"So Mr. Poya," asked Mary, "How long have you worked for Annokkha Drat Exports?"

"I am driver and run errands for people at mine for two years."

"And you live in Kandahar?" Dan asked.

" Baleh—yes," he said as he squirmed a little in the drivers seat. "We should be going. We talk later."

After the brief rest stop, Dan felt energized and started thinking about their upcoming interview. Dan looked to his right at Mary and asked, "Do you have any question about our interview with Abdul Wazir?"

"No, I'm ready."

They resumed their gazes out the window. Poya steered around a long switchback and pulled off the dirt road onto a relatively flat area. He stopped and got out of the driver's side and looked down at the tire. "We have flat tire. Get out while I put on spare tire." he said.

Randal slid out of the right passenger door just as Dan opened his door on the left side. He closed his door and saw a motion in his peripheral vision, and two loud pops rang out, like firecrackers.

Dan instantly perceived danger, and executed a forward karate kick that hit pay dirt, Poya's knee. Poya let out a dull grunt. Dan rose into a defensive stance and chopped his left hand on Poya's arm with such force that it knocked the pistol out of his hand. Dan followed up with his right hand using a forehand punch directly into Poya's nose. Blood erupted and sprayed over the roof of the Ford as Poya fell backwards to the ground. Dan grabbed the gun and looked at Mary through the open window. "Are you all right? Get out and get down out of sight."

In that instant, Dan felt an arm grab him from behind in a chokehold. He felt warm blood drip on his right ear. He smelled the metallic pungency of the blood as it dripped onto his face. Poya's grip around his neck tightened, blocking Dan's airflow.

Dan struggled to widen his stance with his attacker on his back squeezing the life out of his body. Poya was strong. Dan's training kick in. He twisted to his left, lowered his upper body raising Poya's feet off the ground, and used his wide stance to turn his free arm around. He twisted his head free and his right elbow connected on the right side of Poya's face, causing more blood to spatter out of his broken nose. Poya fell to the ground on his face. Dan nearly broke Poya's left arm as he wrenched it all the way up to the backside of his head. Dan heard a dull groan coming from Poya's bloody face.

"Mary, please give me a hand." Dan shouted. "Get me a rope or twine so I can tie him up, and take his gun and aim it at his head. If he moves to get away, shoot him."

With Poya hog-tied, hands to feet, Dan walked around the Explorer to check on Randal. He was lying in a pool of blood with two shots to his head. This was a setup.

Dan reached for the automatic weapon and checked to see if it was loaded. He chambered a round. Then he looked at the tire—it was nearly flat, but he noticed the valve stem cap was gone.

Mary was shaking uncontrollably from the incident as she held the pistol on Poya.

"There may be others out there to help Poya," Dan said. "When you are able to think clearly, let's make a plan." Dan embraced Mary to help her gain control of her fear. "You are fine. You are fine."

She nodded. Her eyes meet his as her face softened with a few sobs and deep breaths.

"Where did you learn how to do that?" she asked. "You saved our lives." Her eyes filled with tears that freely ran down her cheeks.

"Didn't I tell you I'm a black belt in karate? It's hard to describe. My training and instincts kicked in and I reacted as I was trained. This time it was for real. My adrenaline spiked giving me extra strength. It's not over yet—we still need to get back to Kandahar."

"Why do you think that?" asked Mary. "Who could be out here to finish us off?"

"Certainly others are involved, because Poya didn't decide to kill us on his own. He was instructed to do so by someone much higher up the food chain," Dan explained.

"We know the CEO, Rajah Malani, had been playing dirty with the U.S. Lithium Corporation," said Mary. "I wouldn't be surprised if he didn't order this, because he must have something riding on the rare earth metals too."

"We're getting close to learning something we shouldn't. I think it has to do with the rare earth metal gadolinium, which is supposedly mined at the end of this road," Dan said.

"Now it's beginning to make sense," said Mary. "The first two samples that Lizzy received were certainly special, and they don't seem to come from the same source as all the others. What if the mine here has lower quality minerals, and the good mine is somewhere else?"

"That could mean that Abdul, Bibi, and possibly Rajah may be conspiring to get the good metal from this other source and want to keep it secret."

Mary added, "And they are willing to kill to keep it secret. Maybe they want to acquire the other source and make a financial killing. Like Rajah's plan to capitalize on lithium."

"I'll replace the flat tire, then we'll head back to Kandahar and see if the U.S. military will interrogate Mr. Poya. I'll use the satellite phone to call the base to tell them about Sergeant Wilson, and that we are bringing in the killer, Mohammad Poya," Dan said.

"Should you cancel our appointment with Wazir?" Mary asked.

"He may be part of the plan. He probably knows Poya's plan to kill us, so I don't want to let him know we are still alive," Dan said.

"What if Wazir tries to call Poya?" she asked.

"I'll keep his phone near me to see if he does call, but I won't answer it. Maybe Bibi will call him to learn what's happened, so let's keep his phone near us to get any numbers calling in."

Dan retrieved the satellite phone from Sergeant Wilson. He'd never used one before, but attached to the phone were several numbers—one was for the Kandahar base.

"Yes, yes I will," said Dan, after calling the number. The captain took his story and gave him instructions. "Thank you. We will be looking for you."

"Mary, we're going to drive down the hill to the main road into Kandahar. They are dispatching a chopper to escort us to the road where they will pick up Sergeant Wilson's body. Then we will drive to the base while they maintain cover in case we are attacked again."

Three hours later they pulled the Explorer into the security gate at the Kandahar Airport. The guard knew they were coming and directed them to a special security building where they put the hog-tied Poya on a gurney attended by a paramedic. Dan and Mary followed the gurney into the building. They rolled Poya into a small room to treat his injuries, and they met with an Army captain for debriefing.

"We need to know who he's working for," Dan said. "We think there may be a corporate conspiracy within a U.S. corporation, so any information you gather will help us."

"This guy will be handed over to the local authorities after we interrogate and release him. It is their process here in Afghanistan. I will need statements from both of you describing the event," said the Captain. "Justice for murder in this country is quite swift."

"Can you hold him until we know of other conspirators?" Dan asked. "We really need to find out who they are."

AFTER GIVING THEIR STATEMENTS, Dan and Mary returned to the BOQ to shower and eat. They approached Mary's room, which was across and down the hall from Dan's. She hesitated. "Dan, please help me."

"Yes. What is it?" Dan said, but before he reached her door, she grabbed and held him. She was shivering over her entire body as if she were standing naked in a snow bank. Dan held her tight to her body and wrapped his arms around her back. His hug warmed and softened her shivering. She looked up into his eyes. He saw her green eyes with tears running down her cheeks.

"You're going to . . . " She pulled Dan's face to hers, opened her mouth, and welcomed his warm tongue. They stood there for an eternity, embracing, kissing. No words.

She smoothly opened her door and guided their interlocked bodies inside. She looked up at Dan before he could say a word and put her finger on his lips and then guided her lips to his. They stood and enjoyed the embrace. Dan's body was flushed from head to toe. Mary's shivering stopped, and her body was fully engaged with his. She slowly rotated her pelvis to the rhythm of their breathing.

"I was never so scared as I was today," she whispered into his ear. "I won't let you go, ever."

She unbuttoned his shirt. He unbuttoned hers and her bra. Her breasts were beautiful, round and firm. She rubbed his chest, simultaneously guiding him to the side of her bed. She was in control and quickly released his belt and snap and slowly lowered his fly. She felt the warmth and size of his penis and lowered his pants, leaving only his shorts. She sat him on the edge of her bed, removed his shoes and pants, and pushed him back onto the bed. "Relax Dan, you are all mine tonight," she whispered in his ear, nibbled his earlobe, and gently filled his ear with her warm, wet tongue.

Dan watched as she lowered her pants leaving only her fine, lacy panties.

They slipped into bed, and embraced and kissed and explored each other's bodies.

MARY SLEPT LIKE AN INNOCENT ANGEL while Dan tossed and turned, thinking about their next steps. He believed that Rajah, Bibi, and Wazir were involved in a conspiracy to take their lives. Dan was sure that Poya's botched job would not go unnoticed for long. At 5 a.m. he dressed and went to the business center. He checked Poya's cell phone and saw that no incoming calls had been made. He was sure his co-conspirators were alerted because Poya did not report back as planned.

Using a secure Wi-Fi connection, Dan sent Robert Kavanah an email giving him a status update of their situation and asked him to not communicate with anyone at AeroStar, especially the three persons of interest. He also asked him to hire a bodyguard to meet them at the gate in the New Delhi Airport to secure their travel to the Four Seasons hotel in New Delhi. They were booked on the 2:30 p.m. Ariana flight into Delhi at 3:30.

Since Bill's time in New York was 8:30 p.m., Dan also copied the answering service to contact him immediately in this emergency.

After a hot cup of coffee, Dan walked out the back door of the BOQ and down the street toward the security post. At this time in the morning, the air felt cool on his face. His thoughts started to clear up as he thought through yesterday's events. He just couldn't shake the uneasy feeling he had about other players in the big picture.

When he reached the U.S. Army security post, he entered and asked to see the officer on duty. The sergeant at the desk dialed the phone and a few minutes later a captain came through the door. "Good morning, sir. How may I help you?" he asked.

Dan explained who he was and handed him Poya's cell phone. "You may need this to prosecute him for the murder of our bodyguard. But we've been trying to learn who his co-conspirators are. Can you tell me if you've done any background check on him?"

The captain nodded and showed Dan into the next office. "We have a complete dossier on him. His name is Mohammad Poya. He's also known as the Fixer. When the war was at its peak, Poya was used by the U.S. Army to deliver information to the various warring factions in Afghanistan. While he's Muslim, he knows all the Sikhs and Hindus and made good money as a courier of information and messages between the U.S Army and the warlords. Abdul Wazir hired him at the Annokkha Drat mine near here. We were told that he was a courier for him also."

"Do you know if he has any other contacts with the Annokkha Drat people in India or possibly people in the U.S?" Dan asked. "Perhaps you can examine his cell phone to see who else he's contacted. If you do get other names, would you please forward them to me? This is important to us. He's already taken one life." Dan paused. "We think his mission was to kill my partner and me so that the information we gather won't get back to our client."

Dan handed the captain his business card that had both his cell phone number and also his email address. "Please contact me with any information, no matter how trivial it may seem to you."

"I'll see what I can do, but I cannot make any promises."

Back at the BOQ, he quietly entered Mary's room. She was up and packing for the trip to New Delhi. They embraced and kissed. "Dan, I was so worried when I woke up and you were gone."

"I'm here now. You have nothing to worry about."

Mary pulled his belt and whispered in his ear, "Make love to me again."

He needed no coaxing. This time, their lovemaking wasn't rushed. It was slow, smooth, and with the sensual purpose of pleasing each other. Afterwards, her head lay on his shoulder as she drew circles with her finger on his chest. "I'll remember this feeling forever. We nearly died, I know. I hope we'll feel this way tomorrow, the next day, and long after we're done with this client."

Dan rolled his head looking into Mary's eyes, "Yes, we will be together long after this client."

As Mary pulled on her khaki pants and buttoned her shirt, she asked, "Did you find out anything from the Army?"

"Yes. They know all about Mr. Mohammad Poya. He's Muslin and has done a lot of clandestine work for the U.S. Army at the height of the war. He's a bad guy doing bad things. I have a plan to keep us safe until we get home."

"I'm glad to be with you because I know you will keep me safe. Then reality hits me when I look around, and I get nervous in this godforsaken country," she replied. "Thanks for getting security for us in New Delhi. That will help me feel more secure. Do you think they will try to kill us again?"

"I just don't know. But I will do everything I can to keep you safe."

48

New Delhi, India

DAN RECEIVED AN EMAIL that confirmed their new bodyguard. It included his picture so that they would recognize him as the right person. Funny how this worked just like Uber in New York City. Before this technology was available through smartphones, a perpetrator could easily intercept the bodyguard information and access to the subject without their knowledge. It's certainly a best practice to be able to identify the correct person, and also be identified, instead of advertising your arrival with a sheet of paper with "Duggan" written across the front so everyone could see.

As Dan and Mary boarded Ariana 312 in Kandahar, they looked at all of the passengers to get a mental profile of each. Any one of them could be an assassin. After they buckled into their seats, Mary wrote two seat numbers on a piece of paper.

"These two have the characteristics of a person being deceptive. Let's keep an eye on them," she whispered into Dan's ear. He nodded.

"When we land, let's wait and be the last to disembark so we can watch them," Dan said.

"Good idea." Mary squeezed his arm, not wanting to let go. "You think of everything."

When the plane reached altitude, the ride was smooth and they both relaxed, but they didn't close their eyes the entire trip. Mary put her arm through Dan's and laid her head on his shoulder for the entire trip.

Soon, the plane was landing. Mary peeked at the two suspects. "Our guys look okay to me, but let's keep an eye on them when they get off the plane," she whispered.

Mary by the window and Dan in the middle seat, they watched all the passengers disembark. They took their time and slowly made their way to the plane's door and up the gangway that connected to the building. Dan stepped into the passenger area first and looked on both sides of the doors. He recognized their bodyguard sitting in the waiting area near the passageway. With no other people in the area, Dan and Mary stepped up to him and shook hands.

"Thank you for meeting us at the gate," said Dan. "My name is Dan Duggan and this is Mary Johnson." He was a tall, buff American—Ex GI no doubt.

"Pleasure to meet you two. My name is Jay Jackson, and I will be with you until you board your plane to the U.S. tomorrow," he said.

"Where do we go from here?" Mary asked.

"First of all, do you have luggage?"

"No, we just have carry-ons," said Dan. "We left most of our luggage at the hotel."

"That's good because I will take you the back way to my vehicle. Just follow me."

A few feet away, he opened a door marked authorized personnel only.

"This is a secure passageway used by the airline and airport people who have the proper clearance. Because of my work, I also have a badge and know the secure passwords to some of the areas," he said. "It makes my job much easier."

The secure hallways had little traffic of people going to and from their jobs. As they approached each person, they made eye contact and then let the person pass to one side; they kept careful vigilance on all moves. After they walked down stairways, they exited to a parking area.

"Here's where I park. I want you two to stay in this alcove while I clear the area. When I'm ready, I'll wave you to proceed. My vehicle is the black Cadillac SUV on the end over there," he said.

"We will be driving for about forty-five minutes from the airport to the Four Seasons. We'll pass through some poor sections of town—not to worry. This vehicle is specially built for dignitaries and corporate executives who need protection. It has bulletproof glass and armor plating so that anyone trying to assassinate you from the outside will fail.

"The most vulnerable places I'm concerned about are entering and exiting the SUV and buildings. The Four Seasons has a special carport for VIPs to minimize this risk. I will clear the carport and the entryway into the hotel before you open your door. Then I will lead you in and clear the hallway. We'll be using a VIP elevator to your rooms—here are your passkeys. Your stored luggage will already be in your rooms. Do you have any questions?" asked Jay.

"What if we want to go out for dinner or to the business center?" asked Dan.

"I recommend that you eat in your rooms tonight and tomorrow morning for breakfast. When we go to the airport, we'll follow the same procedure."

"Have you ever had a problem—like someone trying to kill your client?" asked Mary.

"Ma'am, I've been attacked numerous times in Kabul, Kandahar, and other Afghani villages. I'm still here. I haven't seen an attack since I've been working in New Delhi. But that doesn't mean I won't follow my strict protocol for security."

Mary sat back and relaxed—she noticed the house with the elephant garage and a few other landmarks she remembered on their route to the hotel. It seemed like a lifetime ago when they passed through this neighborhood—seeing people in a different culture, living and working. She felt a pang of homesickness ant thought of her little apartment in New York City. Her life was so simple, so rewarding, and so wonderful. Now it's changed. Working with a man she'd fallen for.

Her face and body glowed as she thought about Dan. How he saved her life. What will life be like, now—with Dan? She had a lot to think about, but for now she must be sharp and strong to finish the engagement.

Dan kept his eyes on the road ahead of them and to both sides. *What if they try to blow us up with an IED in route to the hotel?* The Cadillac was quiet and smooth with a gentle sway that allowed them to relax into the deep, soft leather.

Dan closed his eyes and thought about his night with Mary. How she was like an addictive elixir that he couldn't get enough of. How her kiss was deliciously wet and warm, like a sip of fine red wine that gave him warmth and euphoria. His dream of her seemed like an eternity until he felt a warm hand squeeze his.

"We're here," said Mary. She unbuckled her seatbelt and slid next to Dan hugging his neck. "I think you were dreaming," she whispered. "I dreamt of you too."

Jay looked back over his seat, "We are in the Four Seasons carport. I'm going to clear the area so we can enter. Do you have any questions?"

We looked at each other, "We'll stay here until you signal clear?"

"Yes, wait until I clear the area and the entry door. Then I'll wave you in from the door, so I can see the whole area."

A few minutes later, Jay waved to them as a signal to exit the SUV. They each grabbed their two small bags and walked toward the entry door just as a bellhop wearing a bright red suit with a colorful top hat pushed open the door from the inside.

Ten feet from the door, Mary saw the man's face and suddenly yelled, "He's the one!"

She pushed Dan as hard as she could to her right, and then dove to her left just as a loud "pop-pop" rang out. Both bullets glanced off the door of the Cadillac as they fell to the ground.

Jay reacted using his skilled hand-to-hand maneuvers to attack and disarm the assassin. When Mary saw the attack was over, she ran to Dan and hugged him, "Are you all right?"

"Yes, ah, I . . . I think so. You saved my life," Dan stammered as he slowly got to his feet.

Meanwhile, several security staff from the hotel rushed out and secured the attacker while others helped Dan and Mary into the entry door.

Mary looked into Dan's eyes with tears running down her cheeks. "He's the same guy I pointed out on the plane. I thought he killed you." Her tears flowed and they just stood there in an embrace.

LATER IN THE EVENING, they ate dinner together in Dan's hotel room. Their mood was somber. They felt extremely lucky—not just once, but twice. And they were falling in love.

"I don't want to talk about business now," said Dan. "But I need to call a confidant about these incidents because they do have something to do with our engagement. I'll go into your room to make the call. Just give me five minutes, okay?"

"No. Stay here with me. I won't listen, promise."

At the desk in his room, Dan dialed a New York telephone number. Mary sat at his side.

"Hello, this is Dan Duggan from Anderson & Smith. Yes, I'll wait." After a short pause Dan continued. "It's extremely important that I tell you what's transpired on our trip to India and Afghanistan. Yes. Yes. We had a productive meeting with Bibi and his team in New Delhi, and we discovered an anomaly with his original ore samples of gadolinium." Dan paused while he listened. "Yes, that's right. We went to Kandahar to visit the gadolinium mine there. Yes. Yes. No. On our way to the mine, our driver ambushed us. It was a setup. He killed our bodyguard, but we were able to disarm him." A pause. "No, we are unharmed. His name is Mohammad Poya and was hired by Abdul Wazir at the mine. Yes." Dan then continued on about the attack at the Four Seasons hotel by a man who followed them onto the plane in Kandahar.

"We think there is a conspiracy within the company. Mary and I believe it could possibly include Rajah Malani, Bibi Shinwari, and Abdul Wazir. Yes. No. Okay."

"We think the motive may be due to the poor quality of gadolinium coming from the Annokkha Drat mine, or some financial incentive with the Aether Program. Yes. Yes we will," Dan said. "Will you look into all communications between the U.S. employees and the Annokkha Drat Exports employees the past few days?

Perhaps someone in headquarters is giving them information about our travel plans. Okay?"

"Yes, I will call you when I get back to New York. Thank you," Dan said and hung up.

"Who was that?" asked Mary.

"I can't tell you right now, but it's someone within AeroStar who's been helping me from the inside. This person can help us," said Dan.

49

Manhattan

"HERE'S TO THE CONCLUSION of an adventurous engagement," said Dan as he held his glass of Tempranillo to Mary's.

"I will toast to that," she said. "But I want to add something. Here's to you, Dan, the man I fell in love with while in Afghanistan."

Clink!

"And to you," Dan said as he held up his glass again. "The woman who fills my heart with joy beyond belief."

Clink!

They swirled their glasses, sniffed, and carefully let the mellow red wine fill their palates. The wine produced a warm glow as it flowed through their bodies.

They ordered Spanish tapas and slowly indulged in good food and good company. Since their return, they had polished the final report and presented it to the Board of Directors. The Board approved the plan.

The incident in Afghanistan was traced to Abdul Wazir and they learned of his firing. Mohammad Poya was tried, convicted, and hanged to death by the local Afghanis.

"It's too bad that we couldn't find a smoking gun that pointed directly to Rajah or Bibi," said Dan.

"Well, it really wasn't our issue to solve anyway. We just stumbled across it," Mary replied and shrugged her shoulders.

"But we did deliver an excellent report and a way forward that the management team will be able to deliver on," said Dan. "And, the bonus we received really helps."

They were about done with their last tapas when Mary's cell phone rang. "Dan, I'm sorry I left the ringer on." She pulled her cell phone from the side pocket in her purse and looked at the caller. "It's from Conrad French. I wonder why he's calling me."

"He probably wants a date," Dan said sarcastically. Suddenly the mood of the evening changed.

Mary hit the on/off button to kill the ringing, "No, I don't think so. We agreed to be good friends and help each other when needed." Her tone was like a third grade teacher telling her students to put crayons away.

"Well, in that case, you should see if he leaves you a message. I bet he's in town and wants to see you."

Mary was pretty confident in herself, especially when it came to reading and communicating with possible suitors. "I'll take that challenge." She retrieved her cell phone and showed Dan a waiting voice message. "Here you go." She hit the voicemail button with the speaker on.

"Hi Mary, it's Conrad. I'm sorry for calling so late, but you know how the three-hour time change always gets me confused. I want to tell you about a major breakthrough in the Kevin Koubiel murder. You know we solved the case, but I have a new finding that may impact your project. Please call me at any time."

"So he's not calling me for a date," said Mary. "This sounds important. Do you want to return his call now?"

"Yes. Definitely."

Mary hit the return call button leaving the speaker on.

Conrad picked up immediately. "Mary thanks for returning my call. It's good to hear from you. How is the progress on the AeroStar engagement?"

"We finished our final report. I have Dan with me. Do you mind?"

"No, you both need to hear this. We solved the Kevin Koubiel murder. You may recall that another mob boss in Las Vegas wanted to muscle in on the lithium business."

"Yes, we thought you arrested the shooter," Mary said.

"We did arrest the shooter and we identified the co-conspirator who masterminded the hit. Do you know Mr. William Anderson? He is the VP of Research and Development at ComStar in St. Louis."

Mary and Dan stopped and just looked at each other mouthing *no way*. "Conrad, this is Dan. Are you sure about this?"

"Absolutely. We learned that he got himself into a little financial problem in Las Vegas. The big boys took over and they made him an offer he couldn't refuse.

"From wiretaps we put on his phone, he was also making a deal with a mine in Afghanistan. One that produces a rare earth metal called gandilumiun, or something like that. The big boys in Las Vegas also wanted in on that one, so they ponied up some muscle and used Mr. Anderson to do the dirty work."

"Do you mean the rare-earth metal called gadolinium?" Dan asked using clear, slow pronunciation.

"Yes, that's the one. Have you heard of it?" asked Conrad.

"Yes, we know it well," said Dan. "How soon are you going to arrest him?"

"I can't tell you that because the FBI will do the deed. Listen, if you know Mr. Anderson, keep this confidential. Okay?"

"Conrad, we know Bill Anderson well. He's a big executive on our project. Should we let someone in the company know about this?" asked Mary.

"No. Absolutely not. This could cause him to run. Only you and Dan know about this, and no one else in your firm or ComStar or AeroStar should know either."

"Tell the FBI that Mary and I were just in India and Afghanistan to investigate anomalies in their gadolinium mine, and two different people tried to kill us," said Dan. "This may have also been the work of our common friend, Mr. Anderson. Mary and I gave statements on both attacks to the police. One attempt was on our way to the mine in Afghanistan, and the second attempt was at the entrance to our hotel in New Delhi. Our U.S. Marine bodyguard in Afghanistan was killed by the gunman."

"I will," said Conrad. "They may call you with questions."

"That's fine with us," said Dan. "Thank you for the information, as this is important to us. We were thinking of another person in the company, but we had no proof of that person's involvement. We just knew it was serious enough to take extra security precautions, and it paid off. Mary, do you have anything to say?"

"Yes, I do," she chimed in. "Conrad, I am really blessed to have the opportunity to work and learn from you. Thank you very much."

"You two are very welcome, and thank you for helping my case. Do you two ever stop working?"

"We're discussing the recommendations in our final report and celebrating with a glass of Tempranillo. With what you told us tonight, the final puzzle piece fits perfectly. We will deliver our final report next week," said Dan.

"Well, enjoy," said Conrad. "Good night."

"Have a great night also," they said together.

50

Teterboro, New Jersey

"LET ME TAKE YOUR BAG," "Is this all you have?"

"Yes, Dan," said Mary. "You know I pack light. Did you forget our little side trip to Kandahar?" Her tone was a little patronizing.

"Okay," he said as he threw both bags into the back seat. "Are you ready for a little trip? Just you and me and maybe some other tourists too?"

"Yes, I'm ready. Why aren't you telling me where we're going?"

"Because it's a surprise."

Being the gentleman he was, Dan opened her door and held her hand as she sat. "Remember to buckle your seatbelt," he reminded her.

"Yes, yes I know."

Dan fiddled with his GPS and opened the window.

"Clear," he yelled loudly, and then looked out all four windows, front and back.

As he turned the ignition, the motor cranked with a distinct high, whining sound; swish, swish, swish followed by a loud Grrrrruppppppppppppppp when the motor roared to life. He quickly checked the gauges to ensure the motor was producing oil pressure, electrical charge, and RPM.

Dan looked over to Mary on his right. She's so beautiful, even wearing a David Clark headset.

"Can you hear me?" asked Dan.

Mary nodded, giving him the thumbs up sign.

"Now speak into the microphone on your headset."

"Yes, I hear you loud and clear. Do you hear me?"

"Yes, I do."

Dan flipped a few more switches and turned some dials writing down notes on the clipboard strapped to his right leg.

"Teterboro Ground, Cessna November 9er Whiskey Tango Foxtrot at First Aviation with ATIS, request taxi to runway 6."

"Cessna 9er Whiskey Tango Foxtrot, clear to taxi runway 6."

This is Mary's second flight with Dan in the Cessna, so she's beginning to understand some of the piloting jargon, and she was totally comfortable flying with Dan. They had planned this trip for at least a month: "Only pack clothes for one day and bring your camera."

But Dan didn't reveal their destination, "It will be a big surprise. You will love it."

After going through the checklist and run-up tests, Dan called the tower for takeoff clearance. As they took off to the northeast, the sky was clear and the leaves were just beginning to turn colors.

Dan called departure control and requested flight following, and then turned the nose of the plane to his compass heading. He climbed to 5,000 feet, high enough to see the beautiful countryside and just below most of the controlled air spaces. With radar flight following, Dan was comfortable that he will be alerted to other traffic in the area. This was a simple pleasure flight with the woman he loved, so he wanted safety without the burden of executing an instrument flight plan.

As they flew, Dan pointed out various landmarks that Mary had never seen before, certainly not from the air. He figured that their flight path to his surprise destination would take about two-and-a-half hours. He was busy keeping the plane on his compass heading and watching for other aircraft.

When they passed over the New Jersey/Pennsylvania border, Dan pointed out they were getting closer.

"I don't know what the big surprise is going to be," exclaimed Mary.

"You'll enjoy this. Not many people get to see this spot from the air."

Dan continued to point out landmarks on their flight path and explained some of the flight and navigation instrument. Mary listened intently as if she would be soloing some time soon.

"The water you see ahead of us is Lake Erie—one of the great lakes, as you know. Then over to the right you see the city of Buffalo coming into view."

"I still don't get it," said Mary.

He pointed out more landmarks like Lake Ontario and Lake Huron "way over there." A few minutes later, he reduced power and descended to one thousand feet. He pointed out another landmark way to their left and turned the plane to the right. He pointed down and straight ahead to a massive amount of water with little islands of green.

Dan pointed ahead, "Look at that."

Straight ahead was a massive rainbow that arched over the entire mass of water. No rain in sight. The air looked like a mass of fog rising up from the water.

They both watched in wonder. Dan banked the plane into a gentle right turn, so Mary could look down from her window.

She gasped. "Dan, this is the most beautiful view I've ever seen in my life."

Directly below her was Niagara Falls. He circled slowly to experience the size and scale of the most beautiful falls in the world. "Thank you for such a wonderful surprise," she said.

A short time later they were on the ground and taxied to the transit area of the Niagara Airport. Mary helped Dan tie down the plane. Uber was waiting for them for the short drive to their hotel.

"Mary, this is to us." Dan held his glass of champagne to Mary's on their balcony as they listened and watched the falls at close range.

"Dan, you are the greatest."

Engagement Research Binder

This binder contains the external research done by Dan Duggan while performing the business strategy review of AeroStar's current strategic initiative.

Advanced Space Transportation Program by NASA: Paving the Highway to Space

Going to Mars, the stars, and beyond requires a vision for the future and innovative technology development to take us there. Scientists and engineers at NASA's Marshall Space Flight Center in Huntsville, Alabama, are paving the highway to space by developing technologies for 21st century space transportation.

As NASA's core technology program for all space transportation, the Advanced Space Transportation Program at the Marshall Center is pushing technologies that will dramatically increase the safety and reliability and reduce the cost of space transportation. Today, it costs $10,000 to put a pound of payload in earth's orbit. NASA's goal is to reduce the cost of getting to space to hundreds of dollars per pound within 25 years and tens of dollars per pound within 40 years.

The high cost of space transportation coupled with unreliability is a virtual padlock on the final frontier. But imagine the possibilities when space transportation becomes safe and affordable for ordinary people. Whether it's living and working in space, exploring new worlds, or just leaving the planet for vacation, the opportunities for business and pleasure on the space frontier are endless.

Our dreams of everyday life in space and its promise for a better life on earth are hostage to the high cost of space transportation. That's why Marshall Center scientists and engineers are pushing a variety of cutting-edge technologies—from simple engines to exotic drives—to reduce the cost of space transportation and open the final frontier.

Next-generation Launch Vehicles

Dramatic improvements are required to make space transportation safer and more affordable. Future space launch vehicles must be safer, more reliable, simpler, and highly reusable. The Advanced Space Transportation Program is developing technologies that target a 100-fold reduction in the cost of getting to space by 2025, lowering the price tag to $100 per pound. As the next step beyond NASA's X-33, X-34, and X-37 flight demonstrators, these advanced technologies would move space transportation closer to an airline style of operations with horizontal takeoffs and landings, quick turnaround times, and small ground support crews.

This third generation of launch vehicles—beyond the space shuttle and "X" planes— depends on a wide variety of cutting-edge technologies, such as advanced propellants that pack more energy into smaller tanks and result in smaller launch vehicles. Advanced thermal protection systems will also be necessary for future launch vehicles because they will fly faster through the atmosphere, resulting in higher structural heating than today's vehicles.

Another emerging technology—intelligent vehicle health management systems—could allow the launch vehicle to determine its own health without human inspection. Sensors embedded in the vehicle could send signals to determine if any damage occurs during flight. Upon landing, the vehicle's onboard computer could download the vehicle's health status to a ground controller's laptop computer, recommend specific maintenance points, or tell the launch site it's ready for the next launch.

Air-breathing Rockets

The Advanced Space Transportation Program is developing technologies for air-breathing rocket engines that could help make future space transportation like today's air travel. In late 1996, the Marshall Center began testing these radical rocket engines. Powered by engines that "breathe" oxygen from the air, the spacecraft would be completely reusable, take off and land at airport runways, and be ready to fly again within days.

An air-breathing engine—or rocket-based, combined cycle engine—gets its initial takeoff power from specially designed rockets, called air-augmented rockets, that boost performance about 15 percent over conventional rockets. When the vehicle's velocity reaches twice the speed of sound, the rockets are turned off, and the engine relies totally on oxygen in the atmosphere to burn the fuel. Once the vehicle's speed increases to about 10 times the speed of sound, the engine converts to a conventional rocket-powered system to propel the vehicle into orbit. Tests continues at General Applied Sciences Laboratory facilities on Long Island, New York.

Information on the Internet

For more information on NASA's Advanced Space Transportation Program, visit Marshall's Space Transportation Directorate website.

SpaceX – The Company

Founded by Elon Musk.

Achievements of SpaceX include the following:

- The first privately funded, liquid-fueled rocket (Falcon 1) to reach orbit (28 September 2008)

- The first privately funded company to successfully launch (by Falcon 9) orbit and recover a spacecraft (Dragon) (9 December 2010)

- The first private company to send a spacecraft (Dragon) to the International Space Station (25 May 2012)

- The first private company to send a satellite into geosynchronous orbit (SES-8, 3 December 2013)

- The landing of a first stage orbital capable rocket (Falcon 9) (22 December 2015 1:40 UTC)

SpaceX Business Goals:

Elon Musk has stated that one of his goals is to improve the cost and reliability of access to space, ultimately by a factor of ten. The company plans in 2004 called for "development of a heavy lift product and even a super-heavy, if there is customer demand" with each size increase resulting in a significant decrease in cost per pound to orbit. CEO Elon Musk said, "I believe $500 per pound ($1,100/kg) or less is very achievable."

A single-stage-to-orbit (or SSTO)

SSTO vehicle reach orbit from the surface of a body without jettisoning hardware, expending only propellants and fluids. The term usually refers to reusable vehicles. No earth-launched SSTO launch vehicles have ever been constructed. To date, orbital launches have been performed either by multi-stage fully or partially expendable rockets, the Space Shuttle having both attributes.

Application to interplanetary Travel

When used to move a spacecraft from orbiting one planet to orbiting another, the situation becomes somewhat more complex, but much less delta-v is required due to the Oberth effect.

For example, consider a spacecraft travelling from the Earth to Mars. At the beginning of its journey, the spacecraft will already have a certain velocity and kinetic energy associated with its orbit around Earth. During the burn the rocket engine applies its delta-v, but the kinetic energy increases as a square law until it is sufficient to escape the planet's gravitational potential, and then burns more so as to gain enough energy to reach the Hohmann transfer orbit (around the sun). Because the rocket engine is able to make use of the initial kinetic energy of the propellant, far less delta-v is required over and above that needed to reach escape velocity, and the optimum situation is when the transfer burn is made at minimum altitude (low periapsis) above the planet.

At the other end, the spacecraft will need a certain velocity to orbit Mars, which will actually be less than the velocity needed to continue orbiting the sun in the transfer orbit, let alone attempting to orbit the sun in a Mars-like orbit. Therefore, the spacecraft will have to decelerate in order for the gravity of Mars to capture it. This capture burn should optimally be done at low altitude to also make best use of the Oberth effect. Therefore, relatively small amounts of thrust at either end of the trip are needed to arrange the transfer compared to the free-space situation.

However, with any Hohmann transfer, the alignment of the two planets in their orbits is crucial—the destination planet and the spacecraft must arrive at the same point in their respective orbits around the sun at the same time. This requirement for alignment gives rise to the concept of launch windows.

Rare Earth Elements Neodymium, Gadolinium, and Samarium

Neodymium, Nd, is as a component in the alloys used to make high-strength magnets. These magnets are widely used in such products as microphones, professional loudspeakers, in-ear headphones, and computer hard disks, where low magnet mass (or volume) or strong magnetic fields are required. Larger neodymium magnets are used in high-power-versus-weight electric motors (for example, in hybrid cars) and generators (for example, aircraft and wind turbine electric generators).

Gadolinium, Gd, metal possesses unusual metallurgic properties, to the extent that as little as 1% gadolinium can significantly improve the workability and resistance to high temperature oxidation of iron, chromium, and related alloys. Gadolinium as a metal or salt has exceptionally high absorption of neutrons and therefore is used for shielding in neutron radiography and in nuclear reactors.

Samarium, Sm, is used in samarium-cobalt magnets, which have permanent magnetization second only to neodymium magnets; however, samarium compounds can withstand significantly higher temperatures, above 700 °C (1292 °F), without losing their magnetic properties.

There are three natural minerals forms containing gadolinium:

1. Gadolinite is the most common ore form that contains the element gadolinium. It is identified by its dark, almost black crust when exposed to air, and the inter compound is pinkish to red. It is the most common mineral form, but also contains the lease amount of gadolinium per ton.

2. Bastnaesite Burundi, another gadolinium mineral, is recognized by its yellowish hue and porous texture. It is the second most common mineral form.

3. Lepersonnite-(Gd)-Studtite-Curite is the least common mineral containing gadolinium. It is the purest form of the mineral and is easily identified by its bright yellow, green, and orange crystals resembling feathery ice crystals.

Hypersonic Spacecraft

HTV-3X – Blackswift

The Blackswift was a proposed aircraft capable of hypersonic flight, designed by the Lockheed Martin Skunk Works, Boeing, and ATK. The USAF states that the "Blackswift flight demonstration vehicle will be powered by a combination turbine engine and ramjet, an all-in-one power plant. The turbine engine accelerates the vehicle to around Mach 3 before the ramjet takes over and boosts the vehicle up to Mach 6.

Rockwell X-30

The X-30 is a single-stage-to-orbit hypersonic space plane designed by Rockwell International. The design was from NASA, but the military adopted the design capable of carrying military payload in low-earth orbit and returning to earth. The project was cancelled in 1993.

North American X-15

The North American X-15 was a hypersonic rocket-powered aircraft operated by the United States Air Force and the National Aeronautics and Space Administration as part of the X-plane series of experimental aircraft. The X-15 set speed and altitude records in the 1960s, reaching the edge of outer space and returning with valuable data used in aircraft and spacecraft design.

Boeing SST Design

The Boeing 2707, an SST design, had a swing-wing (swing-wing was the hot thing during that time).

The engines were carried on the stabilizers. During the development Boeing had significant problems associated with the swing-wing mechanism. Concerns about stability and payload capability led the design to be lengthened and added canard to the front to meet rotation requirements, though the new design 2707-200 kept the swing-wings.

However, these changes did little to solve the problems. When kept near the centerline, the hinge mechanism interfered with the undercarriage. The hinge mechanism was so heavy that it negated the advantages of swing-wing, leading to a very poor payload-to-weight ratio. Studies indicated that the aircraft would run out of fuel halfway across the Atlantic, essentially killing the project.

Due to this, Boeing decided to use a fixed-wing platform in its next iteration, the 2707-300, which looked quite similar to the British Concorde.

Specific Impulse Table

Engine	Effective exhaust velocity (m/s, kg·m/(s·kg))	Specific impulse (s)	Energy per kg of exhaust (MJ/kg)
Turbofan jet engine	29,000	3,000	~0.05
Solid rocket motor	2,500	250	3
Bipropellant liquid rocket motor	4,400	450	9.7
Ion thruster	29,000	3,000	430
Nelson ion thruster	210,000	21,400	22,500

References

"Issue-Based Technique" was developed by SOE Inc., 1996, 2008. Betty Sugarman, Principal, has given the author permission to refer to the technique in this novel.

The engagement research binder consists of information collected from the Internet including Wikipedia, company websites, and various organization websites.

Cover Photograph Acknowledgments

Clockwise from top left:

Apollo 4, Saturn V. National Aeronautic and Space Administration.

X-15 hypersonic, rocket-powered aircraft. U.S. Air Force.

Afghanistan mountains. © iStock photograph.

X-30. National Aeronautic Space Plane (SSTO) proposal.

Acknowledgments

Thanks to:

My wife, Judi, for letting me work so many hours on this project.

My friend, Ron Schultz, for following my progress and reminding me this was "just a hobby."

My "East Coast" friends who gave me good ideas for some of my characters:

> Buddy Capella for his idea for the Mafia connection "Il Capo."
>
> Conrad Fongemie, who gave me insight into how detectives investigate a crime.
>
> Kevin Kossbiel, whose business acumen got him killed early in the book.
>
> Michael Iadarola, who rolled his eyes whenever I talked about rockets and said, "What do I know?"

My yoga instructor, Mary Lundstrom, whose inspiration touched many people, including me.

My editor, David Colin Carr, who took my original manuscript and transformed it into a novel.

My copy editor, Arlene Miller, who pushed it over the top.

My publisher, Waights Taylor, Jr., who published this novel.

About the Author

Bobby Leonard was born in Tampa, Florida, in 1948 to a U.S. Air Force pilot. He was raised in Roswell, New Mexico, during the years his father flew the B-36 Peacemaker that carried a Hydrogen Bomb during the cold war era. Bobby's life was influenced by his father's war stories of flying B-29s over Tokyo and Korea. Bobby received his B.S. in Aeronautical Engineering, and his M.B.A from California Polytechnic State University at San Luis Obispo. Bobby's 40-year career includes experience in aerospace and management consulting.

Company Secrets is Bobby Leonard's first novel, a mystery, includes experiences and life stories from his career as a management consultant, and his love for space travel.

Made in the USA
Columbia, SC
04 March 2022

56974290R00139